Tourism Mc 1(

ON[ ]( ](

Through a fascinat[ ] nge of case studies drawn from across the world, this innovative book shows how places to play are also places in play: made and remade by the mobilities and performances of tourists and workers, images and heritage, the latest fashions and the newest diseases. Criss-crossed by various mobilities and performances, places are shown to 'move' as they are put into play in relation to other places.

Part One focuses on the performance of paradise in the Caribbean, the Mediterranean, in 'ecotourism' sites and on beaches across the world. Part Two turns to the performance of global heritage from quaint English spa-towns and Viking heritage centres, to the spiritual sublime of Inca ruins and the Taj Mahal, and the international travels of art exhibitions. Part Three concerns the remaking of global cities: Hong Kong, Barcelona, Rio de Janeiro, and the 'high-rise' suburbs of post-millennial Europe. The last part takes an unpredictable tour of round-the-world websites, Dubai's Palm Island, and the dark worlds of nuclear tourism and deadly tourism.

Tourism itself is thus in motion, becoming less predictable as changing processes of leisure, authenticity, and risk re-shape global mobilities.

Mimi Sheller is Senior Lecturer in Sociology and **John Urry** is Professor of Sociology at Lancaster University. They co-direct the new Centre for Mobilities Research at Lancaster University.

Tourism Mobilities:
Places to play, places in play

Edited by Mimi Sheller
and John Urry

LONDON AND NEW YORK

First published 2004
by Routledge
11 New Fetter Lane, London EC4P 4EE

Simultaneously published in the USA and Canada
by Routledge
29 West 35th Street, New York, NY 10001

Routledge is an imprint of the Taylor & Francis Group

Typeset in Garamond by
Taylor & Francis Books Ltd
Printed and bound in Great Britain by
TJ International Ltd, Padstow, Cornwall

British Library Cataloguing in Publication Data
A catalogue record for this book is available from the British Library

Library of Congress Cataloging in Publication Data
Tourism mobilities : places to play, places in play / edited by
Mimi Sheller and John Urry.– 1st ed.
        p. cm.
 Includes bibliographical references and index.

 1. Tourism–Environmental aspects. 2. Tourism–Social aspects.
 I. Sheller, Mimi. II. Urry, John.
 G155.A1T68 2004
 338.4'791–dc22

                                    2004000999

ISBN 0-415-33878-6 (hbk)
ISBN 0-415-33879-4 (pbk)

Contents

List of illustrations   viii
Notes on contributors   ix
Preface   xii

1   Places to play, places in play   1
    MIMI SHELLER AND JOHN URRY

PART I
Performing paradise   11

2   Demobilizing and remobilizing Caribbean paradise   13
    MIMI SHELLER

3   Islands in the sun: Cyprus   22
    RICHARD SHARPLEY

4   Ecotourists on the beach   32
    ROSALEEN DUFFY

5   Surfing: global space or dwelling in the waves?   44
    ROB SHIELDS

PART 2
Performances of global heritage   53

6   Little England's global conference centre: Harrogate   55
    VIV CUTHILL

7   Bodies, spirits, and Incas: performing Machu Picchu   67
    ALEXANDRA ARELLANO

8   On the track of the Vikings   78
    JØRGEN OLE BÆRENHOLDT AND MICHAEL HALDRUP

9   Art exhibitions travel the world   90
    CHIA-LING LAI

10 Reconstituting the Taj Mahal: tourist flows and glocalization   103
    TIM EDENSOR

PART 3
Remaking playful places   117

11 The paradox of a tourist centre: Hong Kong
    as a site of play and a place of fear   119
    NGAI-LING SUM WITH MEI-CHI SO

12 Barcelona's games: the Olympics, urban design, and global tourism   131
    MONICA DEGEN

13 Tourists in the concrete desert   143
    TUE HALGREEN

14 *Favela* tours: indistinct and mapless representations
    of the real in Rio de Janeiro   155
    BEATRIZ JAGUARIBE AND KEVIN HETHERINGTON

PART 4
New playful places   167

15 Playing online and between the lines: round-the-world
    websites as virtual places to play   169
    JENNIE GERMANN MOLZ

16 'Let's build a palm island!': playfulness in complex times   181
    MATTIAS JUNEMO

17 Atomica world: the place of nuclear tourism   192
    KATHLEEN SULLIVAN

18 Death in Venice   205
    JOHN URRY

    Bibliography   216
    Index   233

Illustrations

Figures

2.1	Cuban cars	19
3.1	Paphos Harbour, Cyprus	25
6.1	Harrogate International Centre	63
6.2	Royal Baths, Harrogate	63
8.1	Viking Ship Museum map	80
8.2	The wrecks, Viking Ship Hall	80
8.3	Activity Room, Viking Ship Hall	81
8.4	Construction of a replica Viking ship	81
11.1	Annual visitor arrivals in Hong Kong, 1996–2002	122
11.2	Hong Kong Harbour	123
11.3	Hong Kong SARS	127
11.4	Monthly visitor arrivals in Hong Kong, 2003	128
12.1	Barcelona Maremagnum	138
12.2	Barcelona: 'Clos, no to the ludic ghetto'	139
13.1	Brøndby Strand, SW Copenhagen	144
15.1	One of the daily reports on Ramon's *Let me stay for a day* website	176
16.1	The Palm Jumeirah, Dubai	182
17.1	Hiroshima memorial	193
17.2	Bravo 20 plans	201
17.3	Visitors at Bravo 20	202
18.1	Destroyed Hiroshima now a place for tourist pilgrimage	210

Tables

9.1	Top-ten most popular museum exhibitions in the West, 1997	92
9.2	Top-ten most popular museum exhibitions worldwide, 1998	92
9.3	Top-ten most popular museum exhibitions worldwide, 2000	93
9.4	Top-ten most popular museum exhibitions worldwide, 2001	93

Contributors

Alexandra Arellano, Postdoctoral Fellow, Canadian Research Centre for Urban Heritage and Centre for the Study of Arts, Letters and Traditions (CELAT), University of Quebec in Montreal. Her research interests focus on tourism, heritage, and cultural change, with a particular interest in festivals, Latin America, and development.

Jørgen Ole Bærenholdt, Ph.D., Associate Professor and Head of Geography Section, Roskilde University, Denmark. Published three books on coping strategies in the North (with Nils Aarsæther), and *Performing Tourist Places* with M. Haldrup, J. Larsen, and J. Urry (Ashgate, 2004).

Viv Cuthill, part-time Ph.D. student in sociology, Lancaster University, and freelance Tourism Training Consultant. Her research interests centre on the relationship between service cultures and place cultures within various tourist destinations.

Monica Degen, Lecturer in Sociology, Brunel University. Research focuses on sociologies of the senses, spatial politics, and city life. Recently co-authored a special edition of *Space and Culture* on *Spatial Hauntings* (2001) and *50 Key Thinkers on the City* (2004/5, with Kevin Hetherington). Also published in *International Journal of Urban and Regional Research* and *Society and Space*.

Rosaleen Duffy, Lecturer in Politics and International Relations, Lancaster University. Her research interests centre on environmental politics and tourism in the developing world. She is author of *A Trip Too Far: Politics, Ecotourism and Exploitation* and co-author of *Ethics of Tourism Development*.

Tim Edensor teaches Cultural Studies at Staffordshire University. He has written about automobilities, modes of walking, and the film *Braveheart*. He is the author of *Tourists at the Taj* (Routledge), *National Identity, Popular Culture and Everyday Life* (Berg), and *Industrial Ruins: Uses, Order, Materiality and Memory* (Berg).

Jennie Germann Molz, doctoral candidate in Sociology at Lancaster University. Her research interests revolve around travel and tourism, technology and globalization with particular focus on the practices and representations of round-the-world travel.

Michael Haldrup, Associate Professor in Human Geography, Roskilde University, Denmark. He has published on topics relating to place identity, everyday life, and tourism. Recently co-authored *Performing Tourist Places* with J. O. Bærenholdt, J. Larsen and J. Urry (Ashgate, 2004).

Tue Halgreen, M.Sc. (Sociology, Copenhagen), M.A. (Tourism and Leisure, Lancaster). Currently a Research Assistant, Institute of Local Government Studies, Copenhagen.

Kevin Hetherington, Professor of Cultural Sociology, Lancaster University. Published various books including *Badlands of Modernity* (1997) and *Expressions of Identity* (1998). He writes on space and identity; museums and material culture; and consumption. Currently working on a book re-theorizing commodity fetishism called *Capitalism's Eye* (Routledge). Co-editor of the journal *Museum and Society*.

Beatriz Jaguaribe, Associate Professor at the School of Communications, Federal University of Rio de Janeiro. Published *Rio de Janeiro, cartografias simbólicas* (co-edited 1994), *Fins de Século: cidade e cultura no Rio de Janeiro* (1998), and *Mapa do Maravilhoso do Rio de Janeiro* (2001). Her central theme is the relation between literature, media, and urban culture. Forthcoming is *New Realisms: literature and the media in contemporary Brazil*.

Mattias Junemo, doctoral candidate in Sociology at Lancaster University, also lecturer in tourism studies at Baltic Business School, Kalmar, Sweden. Research interests include urban mobilities, global workers and the aestheticization of society, with Dubai as a case site for such studies.

Chia-Ling Lai recently completed her doctorate on 'Museums in Motion' at Lancaster University. Research interests centre on issues of globalization/national identity, museums and public display, feminist culture theories, and mobilities.

Richard Sharpley, Reader in Travel and Tourism Management, Newcastle Business School, Northumbria University. His research interests focus on tourism and development, island tourism, rural tourism, and the sociology of tourism. Published various books and articles in these areas.

Mimi Sheller, Senior Lecturer in Sociology and Co-Director of the Centre for Mobilities Research, Lancaster University. Published *Democracy After Slavery* (2000), *Consuming the Caribbean* (2003), *Uprootings/Regroundings: Questions of Home and Migration* (2003). Articles in *Slavery and Abolition*, *Theory and Society*, *International Journal of Urban and Regional Research*, and *Theory, Culture and Society*. Chair of the Society for Caribbean Studies.

Rob Shields, Professor of Sociology and Anthropology at Carleton University, Ottawa, the founding editor of *Space and Culture* and on the Board of *Theory, Culture and Society*. Research spans cultural studies, urban geography, architecture, and new media spaces. Published on the seaside and other *Places on the Margin*, on *Lefebvre*, *The Virtual*, and edited books on lifestyle and ethics, shopping malls, tourism, and internet culture. He still makes time for the beach.

Mei-Chi So, a journalist in Hong Kong. She completed in 2002 a master's degree in Globalization at Lancaster University, being a student of Ngai-Ling Sum. She has worked for various newspapers including *Ming Pao* and *Sing Tao*.

Kathleen Sullivan, Ph.D., Lancaster University, is a nuclear policy analyst and disarmament educator. Her research interests include temporal implications of nuclear waste, intergenerational responsibility, and nuclear-inspired art. Also an occasional performer in the self-styled tradition of 'atomic cabaret', she lives in New York City.

Ngai-Ling Sum, Lecturer in Politics, Lancaster University. Her most recent publication is *Globalization, Regionalization and Cross-Border Regions* (co-edited with Markus Perkmann: Palgrave, 2002). She has also published in *Economy and Society*, *Urban Studies*, *New Political Economy*, *Capital and Class*, and *Critical Asian Studies*.

John Urry, Professor of Sociology and Co-Director of the Centre for Mobilities Research, Lancaster University. He is the author/co-author of *The Tourist Gaze* (1990, 2002), *Economies of Signs and Space* (1994), *Consuming Places* (1995), *Sociology Beyond Societies* (2000), *Global Complexity* (2003), and *Performing Tourist Places* (2004).

Preface

All contributors to this book are currently, or have recently been, post-graduate students, staff, or visitors working at Lancaster University. From this small node in the North West of England many of us have been more and more fascinated by extraordinary changes taking 'place' across the globe. We have tried to capture some of these transformations in this book, especially relating to the production and consumption of various places to play, places that are contingently in play through various tourism mobilities.

In developing this collective book the editors are immensely grateful to its contributors, who have made its performance extremely pleasant – even playful. We are also grateful to other colleagues at Lancaster who have made contributions to this book: Sara Ahmed, Anne-Marie Fortier, Sarah Franklin, Bob Jessop, John Law, Nayanika Mukherjee, Celia Roberts, Jackie Stacey, Bron Szerszynski, Divya Tolia-Kelly. Very many thanks also to Joann Bowker and Pennie Drinkall as administrators of the Centre for Mobilities Research, recently established at Lancaster.

Lancaster,
January 2004

Chapter 1

Places to play, places in play

Mimi Sheller and John Urry

Introduction

This book is about how places to play are also places in play: made and remade by the mobilities and performances of tourists and workers, images and heritage, the latest fashions and the newest diseases. The playfulness of place is in part about the urge to travel elsewhere, the pleasure of immersing oneself in another environment, and the fascination with little differences in the materiality of the world. What is it that provokes a fascination for other places? What makes a place desirable to visit? And even if we desire that other place, why do we actually go through the trouble of physically visiting it? Why be in another place? What are the pleasures it can give that are only available through our physical presence?

But beyond that, the playfulness of place is also about the ways in which places themselves are always 'on the move'. Places are 'performed', often on a kind of global stage, and in these performances they are put into play in relation to other places, becoming more or less desirable, more or less visited. Places consist of physical stuff, which is itself always in motion: new hotel developments, airports and roads, eroding beaches and erupting volcanoes, stinging mosquitoes and deadly viruses – a place can bite back. So we also ask: what needs to be mobilized in order to make a particular kind of place, such as a beach, a city, or a heritage site? How much are places being remade in order to draw in and to capture people on the move? If we are elsewhere with many other people, what does this do to those places? How are places to play stabilized and assembled, and how are they undone and surpassed?

We refer to 'tourism mobilities', then, not simply in order to state the obvious (that tourism is a form of mobility), but to highlight that many different mobilities inform tourism, shape the places where tourism is performed, and drive the making and unmaking of tourist destinations. Mobilities of people and objects, airplanes and suit-cases, plants and animals, images and brands, data systems and satellites, all go into 'doing' tourism. Tourism also concerns the relational mobilizations of memories and performances, gendered and racialized bodies, emotions and atmospheres. Places have multiple contested meanings that often produce disruptions and disjunctures. Tourism mobilities involve complex combinations of movement and stillness, realities and fantasies, play and work.

In this introductory chapter we begin by briefly describing contemporary tourism mobilities, mobilities that increasingly constitute global relationships generally. We then set out our approach to examining places to play and places in play, an approach that looks at diverse systems that contingently orchestrate, organize, and perform these 'urges' to be elsewhere and these materializations of places that become the objects of desire. We outline some themes that are developed in the chapters that follow, dealing with places to play in Central and South America, Australia, India, East Asia, the Caribbean, the Mediterranean, and the north Atlantic, as well as 'round-the-world' travel. But first let us preview the structure of the book.

Part 1, 'Performing paradise', focuses on the beach and the island as particular sites of play. In the iconography of contemporary tourism beaches are the global paradise, places of leisure, nakedness, pleasure, and exhibitionism (Corbin 1994; Löfgren 1999).[1] The Caribbean especially has come to signify a place of playful sensuality that sets the 'Edenic' standard for beaches the world over (chapter 2); even in a way for supposed 'eco-tourists' whose desire to get 'close to nature' leads places to be transformed and sands shifted for their play (chapter 4). But we also show how there are other beaches, beaches of movement, where surfers dwell in a shelter of moving water, the tube, a dynamic place of transcendence where surfers and their soundtracks play (chapter 5). And we examine the rise and fall of beach destinations like Ayia Napa in Cyprus (chapter 3), suggesting how entire islands sway to the fickle flows of tourism.

Part 2, 'Performances of global heritage', explores how heritage is mobilized within and through tourism, staged and performed in museums, World Heritage Sites, and 'historic' places. The invention of tradition, heritage, and cultural memory always has 'real' effects: rebranding a place often occurs through a material 'face-lift' in which entire areas are rebuilt, restructured and renamed, often as 'heritage' sites (chapters 6, 7, and 10). Global flows of tourism and capital touch down in local places, transforming localities yet also being subjected to 'glocalization' effects and divergent temporalities (chapters 6, 9, and 10). And in museums, cultures and bodies are put into play as transnational connections are made and un-made, and distinctions (of class, gender, and national 'habitus') worked and reworked through time and space (chapters 8 and 9).

In Part 3, 'Remaking playful places', the contributions turn to the remaking of global cities as key sites where the mobilities of tourism are unleashed. On the one hand we explore how cities like Hong Kong and Barcelona refashion their built environments and their inter-textual signification to perform as 'attractions' on a highly competitive global stage of 'world-class' destinations (chapters 11 and 12). On the other hand we also examine the far more devalued urban landscapes of high-rise suburbs of European poverty (chapter 13) and the semi-mythic *favelas* of Rio de Janeiro (chapter 14) as places of danger and enthralment, monotony, and awesomeness. These are the new places of play at least for a kind of 'postmodern middle class' both fascinated and repelled by their indescribable, indistinct, yet atmospheric post-apocalyptic urbanism.

In the final part, 'New playful places', we challenge the conventional boundaries of tourism studies by turning to unexpected sites for tourist mobilities: people documenting their global travels for friends at home on round-the-world websites (chapter 15), man-made platforms (chapter 16), nuclear reactions (chapter 17), and places of death and decay (chapter 18). The meaning of place is itself in play in a world of risks, simulations, ever-mutating diseases, and terrorism. The final chapters of this book bring out how not all play is simply fun: places to play are often also places of disease, danger, and death, especially as the world is currently haunted both by global terrorism and global tourism. Tourism itself is thus in motion, becoming less predictable as changing notions of leisure, authenticity, and risk dramatically reshape global mobilities.

Tourism mobilities

'Travel and tourism' is the largest industry in the world, accounting for 11.7 per cent of world GDP, 8 per cent of world export earnings, and 8 per cent of employment. This mobility affects almost everywhere, with the World Tourism Organization publishing tourism statistics for over 180 countries (WTO 2002). Almost no countries are not significant senders and receivers of visitors. Internationally there are over 700 million legal passenger arrivals each year (compared with 25 million in 1950) with a predicted 1 billion by 2010; there are 4 million air passengers each day; at any one time 300,000 passengers are in flight above the United States, equivalent to a substantial city; one-half of British adults took a flight during 2001; globally there is one car for every 8.6 people, with a predicted threefold increase in car travel between 1990 and 2030. Such mobility is highly significant for the global environment, accounting for one-third of total CO_2 emissions (see Urry 2003a, for data sources), and contributing to the rising sea levels that are already threatening to submerge many current places to play.

Many places are being put *into* play due to the increasingly global character of these contemporary mobilities. The 1990s have seen remarkable 'time-space compression' as people across the globe have been brought 'closer' through various technologies. There is an apparent 'death of distance' in what is sometimes described as a fluid and speeded-up 'liquid modernity' (Bauman 2000). There are nearly 1 billion internet users worldwide, there are 1 billion TV sets, and new mobile phones are more common than landline phones. But this compression of time-space also stems from rapid flows of travellers, migrants, and tourists physically moving from place to place, from time to time. The scale of physical travel is immense, the largest ever 'peaceful' movement of people across national borders. Almost all societies across the globe are transformed by flows of tourists, as place after place is reconfigured as recipients of such flows. This is not to say that such flows are 'free', uncontained, or undirected; borders, gatekeepers, police, and security guards filter out 'legitimate' tourists, with their tourist visas, from the 31 million refugees also at large in the world today whose arrivals, and detention, in new places are often highly contested.

These global mobilities presuppose the growth of 'tourism reflexivity', a system of governmentality that ensures that increasing numbers of places around the world monitor, evaluate, and develop their 'tourism potential'. This reflexivity involves identifying a place's location within the contours of geography, history, and culture that swirl around the globe, and locating their actual and potential material *and* semiotic resources. A good example of this is Havana's strange legacy of 1950s American cars and their metonymic capacity to signify 1950s American cultural imperialism in its pre-communist heyday. Reflexivity involves networks of systematic, regularized, and evaluative procedures that monitor, modify, and maximize each place's location within this turbulent global order. These networks 'invent', produce, market, and circulate, especially through global TV and the internet, redesigned (and re-redesigned) 'places' and signs of their meaning. 'Tourism reflexivity' involves the institutionalization of tourism studies (such as books like this one), as well as the development of many consultancy and design firms interlinked with local, national and international states, companies, voluntary associations, and NGOs.[2] Almost everywhere, it seems, can be subject to signing and re-designing, even places such as Las Vegas that seem in the past to have had a very fixed 'placing' upon the global stage (see Rothman 2002).

Indeed more generally, there is an incredible range of employment now found within global tourism. Most people across the globe cannot fail to be implicated within, or affected by, these circuits of tourism and travel. Such employment includes travel agencies; transportation; hospitality; bars, clubs, restaurants, and cafés; architecture, design, and consultancy; media to circulate images through print, TV, news, and the internet; arts and sports events and festivals; and NGO campaigns for and against tourist developments. The growth of the tourism industry also more widely reshapes patterns of urbanization, of infrastructure development (roads, airports, ports), of agriculture and food importation, of cultural production and performance, with implications for almost every economic sector. Such cities provide, according to Rothman, the vision of the twenty-first century as shown in the recent re-placing of Las Vegas as a family entertainment centre (2002).

If tourism is transforming the materiality of many 'real' places, it is also having a deep impact on the creation of virtual realities and fantasized places. There are enormously powerful and ubiquitous global brands or logos that increasingly feature tourist sites/sights as key components of the global culture that their brand speaks to and enhances. Corporations over the last two decades have shifted from manufacturing products to producing brands, involving enormous marketing, design, sponsorship, public relations, and advertising expenditure. These brand companies include many in travel and leisure: Disney, Hilton, Nike, Gap, Easyjet, Body Shop, Virgin, Club Med, Starbucks, Coca Cola, and so on. These brands produce 'concepts' or 'life-styles': 'liberated from the real-world burdens of stores and product manufacturing, these brands are free to soar, less as the dissemination of goods and services than as collective hallucinations' (Klein 2000: 22). These lifestyle concepts revolve around generic types of places to play: the hotel pool, the waterside café/restaurant, the cosmopolitan city, the hotel buffet, the theme park, the cocktail, the club, the airport lounge, the bronzed tan, exotic dancing, and the global beach (see Löfgren 1999, on the last; and Rothman 2002, on Coca-Cola's role in the re-placing of Las Vegas). Virtual places to play can also be built from scratch (chapter 16), built from memories (chapter 8), or built on the internet (chapter 15).

If tourism sometimes feels 'unreal', it nevertheless still mobilizes bodies in sensuous encounters with the physical world. Physical challenges such as surfing (chapter 5), hiking to Machu Picchu (chapter 7), becoming a Viking shipbuilder and sailor (chapter 8), or surviving urban 'zones of risk' (chapters 13 and 14) allow the tourist to put his or her own body into play. As Löfgren writes:

> The grammar of landscape experiences includes all the different tourist forms of 'taking in a landscape': to traverse it, pass through it or past it, to dwell in it, sense it, be part of it…landscapes are produced by movement, both of the senses and of the body.
>
> (2004: 106)

Places are not simply encountered, then, but are performed through embodied play. This playfulness requires equipment for the body: hiking shoes and backpacks, discreet cameras and hidden money belts, bathing suits and towels for more or less naked sun-worship, or suitable outfits for dancing. It also requires equipping of the place: signs and walkways, railways and armoured cars, replica historical heritage and postmodern iconic buildings. These micro-mobilities in the staging of tourism are also crucial to the complex play of tourism mobilities across the world.

Such complex mobilities also entail many risks, and part of the reflexivity of tourism is the awareness of the risks it involves. Tourism has always been accompanied by risks to the person – from disease, from attack, from unwanted sexual attention, from travelling through dangerous places where one may be arrested or imprisoned, or from getting lost and being unable to communicate. But today there are also other kinds of risks, associated with new and unpredictable outbreaks of diseases such as SARS (chapter 11) that show exceptional 'mobility' in passing from person to person, shocking terrorist attacks on prime tourist destinations (chapter 18), or in a more mundane sense the risks of moving priceless art collections for international exhibitions (chapter 9) or of choosing to be a tourist in dangerous 'risky' places (chapters 2, 13, and 14). Risks themselves are mobile, such as that of nuclear radiation (see chapter 17), and the mobilities of tourism generate many new risks and resulting 'frisks'.

There are thus countless ways in which huge numbers of people and places get caught up within the swirling vortex of these systems of tourism mobilities. There are not two separate entities, 'tourism' and 'mobilities', bearing some external connection with each other. Rather they are part and parcel of the same set of complex and inter-connected systems, each producing the other. There is a proliferation of countless discourses, forms, and embodiments of tourist places and tourist performances. Tourism mobilities contribute to the production of more-ordered and less-ordered spaces, using generic patterns to shape spaces with certain kinds of affordances (chapter 10), and these orderings are now found everywhere. Indeed in a way there is less 'tourism' that takes place within very distinct kinds of time-space; there is the 'end of tourism' *per se* within a general 'economy of signs'. There are increasing similarities between behaviour at 'home' and 'away'. Tourist sites proliferate across the globe, while everyday sites of activity are redesigned in 'tourist' mode, as with themed shopping, leisure, and work environments.

We might thus say that Mickey Mouse has escaped from his special kingdom and now seems to occupy almost every place across the globe – there are an extraordinary number of global places for Mickey to play in. How then to analyse such places? What does this book offer to make sense of such ubiquitous places to play?

Places to play

The chapters that follow provide a different way of thinking about tourism and mobil-ities. Conventionally there are presumed to be 'tourists' or 'potential tourists' on the one hand, and 'places' that constitute potential 'tourist destinations' on the other. The tourism researcher then examines those forces that propel individual tourists or groups of tourists to travel to one or other such destination place. Psychologists or economists normally see the 'factors' that induce such travel as individual, while sociologists or anthropologists look for more social factors. But in either case there is an ontology of separate events, where places and indeed cultures are presumed to be relatively fixed and given, and that happen to push or pull discrete tourists from place to place and from time to time (see critiques in Coleman and Crang 2002a: 11; Minca and Oakes 2004a and b).

This book by contrast presumes that tourist activities are not so separate from the places that are visited. Those places are moreover not fixed and unchanging but depend in part upon what happens to be practised within them. Moreover, we presume that there are many overlaps between 'tourism' and other kinds of business, professional, and migratory movement, including the businesses of culture and heritage (chapters 6, 8, and 9), sports (chapters 5 and 12), and generating power (chapter 17). Tourism is not some 'pure' leisure activity only conducted within Mickey Mouse's special kingdom.

And the places to play involve performances by various kinds of 'host', and especially by 'guests'. This means that the chapters reject the ontology of separate events and replace this by the analysis of complex systems of diverse intersecting mobilities. These systems involve networks of 'hosts-guests-time-space-cultures' that stabilize certain places as 'places to play', but only contingently. This book is a manifesto for de-centring tourist studies away from tourists and onto networks or systems that contingently produce and re-produce places to play.

Places are thus viewed as economically, politically, and culturally produced through the multiple networked mobilities of capital, persons, objects, signs, and information (especially via various globalizing media). Such mobilities move at rapid yet uneven speed, crossing many borders. One of the first of these places historically was not 'in the west' but 'of the west', namely, the paradise islands of the Caribbean (Sheller 2003a, and chapter 2). It is out of these complex movements that certain places to play are assembled in unpredictable and temporary configurations. Places are not fixed or given but are themselves 'in play' in relationship to multiple mobilities (Rojek and Urry 1997; Ringer 1998; Coleman and Crang 2002b).

Places are thus dynamic, 'places of movement' (Hetherington 1997). They are like ships, moving around and not necessarily staying in one location. They travel, slow or fast, greater or shorter distances, within networks of human and non-human agents. Places are about relationships, about the placing of peoples, materials, images, and the systems of difference that they perform. In particular, places are located in relation to material environments and objects as well as to human meanings and interactions (Coleman and Crang 2002a: 11). Various chapters examine the role of objects that contingently fix certain networks of play and pleasure: surfboards at Bondi and their subsequent development through military technologies (chapter 5), mineral water, teas, and coffees in Harrogate (chapter 6), remnants of Viking boats found in Roskilde fjord (chapter 8), art objects and paintings in various museums and galleries in East and West (chapter 9), massive man-made islands being built in Dubai on such a scale that their carefully engineered shape is only visible from the air or from space (chapter 16), and invisible nuclear radiation in nuclear tourist sites (chapter 17). Recently it has been shown how complexly linked are forms of 'eating and drinking' and the physical surroundings in which these get performed. With the proliferation of multiple kinds of 'food tourism' (Hall *et al.* 2003), there are many different 'eating architectures' (Horwitz and Singley 2004).

Mobility systems only contingently fix places as fit for play (see Judd and Fainstein 1999 for a political economy of places). New places based upon new performances are always 'around the corner', surfacing out of the swirl of economic, social, and cultural processes. These processes swirl around, changing the fixing of place and bringing unexpected new places into play, such as the previously more-or-less inaccessible Machu Picchu, in the Peruvian Andes (chapter 7), while others rapidly go out of play (chapter 3). And as real and virtual places are interleaved with each other the status of 'place' itself comes into question. How can we enter into a 'mapless city' (chapter 14)? Can a round-the-world website be a global place (chapter 15)? What kind of place to play is a nuclear reactor or an imaginary site commemorating nuclear holocaust (chapter 17)?

This contingency of place partly derives from how places to play require performances by hosts *and* guests (see Coleman and Crang 2002b). This is not simply a question of good or bad service. Rather there is a complex relationship between sets of hosts, sets of guests, and the mannered performances that are possible – and especially those that are profitable. Zukin (2003) has examined certain complexities in such performances by analysing the potential purchase of a jug. In Manhattan she performs *flâneurie* in an area of shops selling imported high-quality European crafts; she visually consumes a particular style of jug. In Tuscany she encounters dozens of this 'same' jug as tourist kitsch, while performing as a *tourist* consumer. Later she builds up expertise in this style of jug, so she develops the performance of a *connoisseur*, enjoying the thrill of travel, search, and acquisition. And finally when she 'travels' to eBay she becomes a *commercial* buyer and seller of these jugs. Thus the apparently simple task of buying a jug is performed and perform-able in strikingly different ways, through *flâneurie*, tourist consumerism, connoisseurship, and virtual commercial travelling.

In the same way, both 'hosts' and 'guests' within tourism are engaged in multiple performances as their bodies move through dwelling-places, airport lounges, beaches, bars, restaurants, cities, or museums, and as they communicate with each other via embodied gestures, written texts, translators, telephones, internet connections, and credit-card swipe machines. These embodied performances involve substantial simplifications of the complex cultures they bring into contact, especially so that they can be commodifed. We see how this occurs in the case of ancient Inca and Viking cultures consumed in Disneyesque packages (chapters 7 and 8) or in the cross-class encounters of middle-class tourism to developing countries (chapters 2, 10, and 14). MacCannell explores these complex relationships in the case of 'cannibal tours' in parts of Papua New Guinea (1992; and see Smith 1989). The development of *favela* tours shows something similar as the 'real Rio' gets authentically 'staged' for visitors on safari (chapter 14). And sharp disjunctures between tourists and hosts are 'naturalized' as an aspect of the local culture, in which embodied encounters are carefully staged in manicured environments (chapters 2 and 10).

Places are thus brought into being through systems of organized and/or informal tourist performances and 'host' performances. Tourist performances include *flâneurie*, climbing, collecting, reminiscing, strolling, shopping, talking, drinking, sunbathing, photographing, reading, dancing, posting information on web sites, surfing, eating, driving, and so on. Host performances include guiding, cleaning, cooking, serving, driving, dancing, playing music, selling, smiling, and so on. In later chapters it is shown how tourists are interpellated as Viking boat builders (chapter 8), as devout visitors reading Koranic texts at the Taj Mahal (chapter 10), as consumers at capitalist temples in Hong Kong or Dubai (chapters 11 and 16), and as visitors who take the entire globe as their destination (chapter 15). But in many cases such roles and their staging are also deeply contested, especially when local political struggles impinge on the decision-making processes through which places are reshaped and mobilized, such as the battles over access to the Taj Mahal, the regeneration of Barcelona, or the packaging of Hong Kong.

But if the performances by hosts and guests no longer happen then the place stops 'happening'. Sites with other configurations of material and semiotic resources will rapidly surpass places that no longer afford possibilities for the right performances on an appropriate scale. This is shown below in the case of Cyprus, which recently became too dependent upon the remarkably rapid changing patterns of youth play-making and related travel (chapter 3), and in Hong Kong, whose central role in the Chinese diaspora placed it in the eye of the storm of the SARS epidemic, which led to a cataclysmic fall in visitors (chapter 11). Insofar as places are constructed as 'playscapes' with 'aesthetic coatings' (chapter 16), those that are able to re-coat themselves quickly will best be able to stay in play, but only just.

Significantly, many 'hosts' are increasingly also from elsewhere; they are not immobile objects of the tourist gaze but are often agents, themselves in motion. Many performers of tourist services – especially in major cities – are on the move, passing through that place, and are no more local than 'real' visitors. Many have moved to work in the industries that service tourism, but they too may be engaged in consuming that place at the same time as they perform that place. We see elements of this in Dubai (chapter 16) and especially in the 'urban nightscapes' of Cyprus, Harrogate, and Barcelona (chapters 3, 6, and 12; see Chatterton and Hollands 2003).

Places in play

Places we have seen are situated at different stages and locations within global flows – there are places that go with the flow and those that are left with a spatial fixity of a no-longer 'cool' infrastructure of suddenly unfashionable hotels, transport systems, and so on. Some such places to play 'move' closer to various global centres (which are themselves in play), while others move further away. The very constitution of 'global centres' and 'peripheral places' (and the globalization of some locales as 'global places') is a function of these shifting configurations of mobilities and immobilities.

New places to play are just around the corner, bringing 'places on the margin' into the global order, such as Sellafield in Cumbria, north-west England, as well as other nuclear sites (chapter 17), Parisian suburbs (chapter 13), the globe itself (chapter 15). Beaches may be moved from one place to another to create eco-tourist sites (chapter 4), and entire islands are built from scratch (chapter 16). Other places move further away, as in the case of Cyprus after the Ayia Napa bubble burst (chapter 3) or Hong Kong after SARS led to its pathologization (chapter 11). Such mobilities of place are highly unpredictable. Terrorist bombs can drive tourism away from places like Luxor in Egypt, Kuta Beach in Bali, or Jerusalem in Israel, but even places of high risk and death may be made into desirable places to play (chapter 18).

More generally, a global stage is emerging, bringing the curtain up on new places and experiences for play. Upon that stage, towns, cities, islands, and countries appear, compete, mobilize themselves as spectacle, develop their brand and attract visitors, related businesses, and 'status'. Increasingly, good places to play are often good places to work, as in the case of various places of the good life along the southern Californian coast (Nevarez 2002). However, if those coming to play are only consuming the place as a noisy, disruptive, and carnivalesque 'party nightscape' (as in Harrogate, chapter 6, or Barcelona, chapter 12), then there will be massive conflicts with those who live in such city centres. Urban protests increasingly occur between those occupying the dayscapes and the nightscapes of bustling compressed city centres.

We also see how placement upon the global stage partly operates through mega-events such as the Olympics, World Cups, blockbuster art exhibitions, City of Culture competitions, festivals, and Expos. The identity of place depends upon its location within, and upon, this global stage. Such events cannot be missed, producing enormous travel at specific moments to catch the event live at the right time. Global events involve moments of global condensation, the intense localization of events within unique places due to the fact that they stage unique events. These places 'transform themselves from being mundane places into being special "host city" sites', occupying a new distinct niche within global flows (Roche 2000: 224). Examples examined below include the 'catapulting' of Barcelona onto the global stage with the 1992 Olympics (chapter 12) and the way that international travelling art exhibitions promoted Taipei, the capital of Taiwan, as a place able to send and receive 'globally' important art exhibitions and objects (chapter 9). More generally, the growth of the internet has speeded up the process of making and remaking global centres of this sort. Going around the world becomes an event in and of itself (see chapter 15 on 'round-the-world' websites).

We examine in this book many different kinds of place to play, from all points of the compass. Becoming a global place to play can enable places to enter the global order contingently (or re-enter, as Havana did in the 1990s). We examine various examples of this entry below: the marketing of Cyprus in the wake of its military division (chapter 3); the city of Barcelona, neglected during the dark years of the Franco period (chapter 12); the Islamic oil state of Dubai, which is now the ultimate platform and playful city (chapter 16).

Further, large increases in tourist numbers emanate from many, very different, countries, although of course voluntary travel is overwhelmingly still a privilege of those who are, in global terms, wealthy. But the Orient, which was once visited and consumed by those from the West, increasingly produces large numbers of visitors keen to see for themselves places of the West that appear to have defined global culture, or at least to import western culture in the form of international travelling exhibitions (chapter 9). Hong Kong, we will see, plays a pivotal role in providing a place in play in relationship to East and West, and increasingly for visitors from the rest of China (chapter 11).

Moreover, places to play are places to meet, places where global networks congregate and socialize. Such 'conference tourism' has involved remaking Harrogate from a genteel English town into a lively site of meetings (chapter 6). Surveys of businesses generally show that 'urban sociability' is the key determinant in the location and working of many business services. Research on Hong Kong and Singapore's financial sectors shows their intensive use of bars, restaurants, nightclubs, sporting clubs, parties, sponsored events, and business associations (chapter 11). 'Sociability in these places is the basis on which contacts and opportunities are made, trust and reliability are tested, knowledge and jobs are exchanged, business deals are tested and sealed, reputations are tracked and business is made sociable...the sites of pleasure in the city come alive as business institutions' (Amin and Thrift 2002: 75).

And out of the swirl of global processes new cool places for each new cool generation get produced. Performances of play seem increasingly generation-specific. There is therefore no single global stage but diverse stages. Examples of cool youthful places to play on the global stage at the time of writing include surfing beaches (chapter 5),

Barcelona (chapter 12), and Ayia Napa (see chapter 3, on the complexities of this 'destination'). Other places to play for young people include those with new distinctions of taste focused upon a 'personal' spirituality, as for some visiting Machu Picchu in the Andes (chapter 7) and those going to Belize for 'worthy' eco-tourism through which they distinguish themselves from mere 'tourists' off to Cancun (chapter 4). And some round-the-world travellers create their own youthful global stage through their web pages, performing the global both in travelling, in recounting their travels, and enrolling an audience into their current and future itineraries (chapter 15).

Thus many of these global places to play are hugely complex places where there is no single 'experience'. The governance of such places is immensely complicated, partly stemming from the diversity of flows of people as tourists, migrants, and workers, from both the developed and the developing world. And they are also complex because of the different age, class, gender, racial, and ethnic performances that occur – often within the same highly localized space, as we see below in the case of the Caribbean (chapter 2), the Taj Mahal (chapter 10, and see Edensor 1998), Harrogate (chapter 6), and Ayia Napa (chapter 3).

Places to play are thus often deadly serious for governments, corporations, locals, migrants, and NGOs, since they are endlessly *in* play in relation to multiple, intersecting, and unpredictable mobilities, as we will now see from places around the globe.

Notes

1   Beaches increasingly feature on the cover of 'academic' books; see Mowforth and Munt 2003, and Sheller 2003, for example.
2   This growth of a tourist discourse is amusingly shown in the character of Rupert Sheldrake, an anthropologist of tourism, who is excruciatingly portrayed in David Lodge's fictional examination of global tourism in *Paradise News* (1991).

Part I

Performing paradise

Demobilizing and remobilizing Caribbean paradise

Mimi Sheller

Introduction

The Caribbean has been repeatedly imagined and narrated as a tropical paradise in which the land, plants, resources, bodies, and cultures of its inhabitants are open to be invaded, occupied, bought, moved, used, viewed, and consumed in various ways. How did it become perhaps *the* classic 'place to play' in the global tourism economy? The earliest European representations of Caribbean landscapes conjured up an earthly Paradise grounded in utopian fantasies of tropical fecundity and sustenance without labour (Sheller 2003a; Strachan 2002). From the seventeenth century onwards 'the tropical environment was increasingly utilised as the symbolic location for the idealised landscapes and aspirations of the Western imagination' (Grove 1995: 3). Such accounts drew on a range of precedents, such as the biblical Garden of Eden, the classical garden of the Hesperides, and the Renaissance botanical garden. Islands have a particular resonance in the imagining of paradise, due to their intangible qualities of separateness, exclusivity, and holism, sometimes described as an 'island logic' in European thought (Greene 2000: 140; and see chapter 3). 'It is the fortune, and the misfortune, of the Caribbean,' argues Polly Pattullo, 'to conjure up the idea of "heaven on earth" or "a little bit of Paradise" in the collective European imagination...a Garden of Eden before the Fall' (1996: 142).

The Caribbean island thus became one of the first 'global icons' (Franklin *et al.* 2000) to encapsulate the dreams and contexts of modernity. In an endless simulacrum, earlier literary and visual representations of the 'Paradise Isles' have been mapped into the collective tourist unconscious before they have ever set foot there. The real Caribbean is always a performance of the vivid Caribbean of the imagination. Verdant forests, exotic flora, and tropical greenery serve as powerful symbols of the 'Eden' that is imagined before European (and African) intrusion into the New World. Tobago, for example, is today promoted as a place where you can 'see the islands as Columbus first saw them', while Dominica is described as 'still the primitive garden that Columbus first sighted in 1493, an area of tropical rainforests, flowers of incredible beauty and animals that exist nowhere else in the world' (Noble Caledonian 2002). The Caribbean has been renaturalized as a virgin paradise in ways that continue to inform its contemporary desirability as the ultimate 'place to play' (Sheller 2003a, 2004). It is brought to the consumer in texts, images, and signs, and is fantastically consumed by tourists who draw on existing visual and sensuous performances to make certain kinds of movements in and through the Caribbean viable. As Strachan argues, the 'tourist-consumer appears to buy paradise in the travel agency or airline ticket office or over the phone, but, actually, this prospective traveler has purchased only the promise of "paradise": the collection of ideas, the myth, the electronic or printed messages that has been exported to her' (Strachan 2002: 113).

This chapter considers the many ways in which the contemporary Caribbean is both a place of mobility and a place on the move. Through a complex historical approach to tourism's intersections with many other mobilities, I argue that it is not only people and things that move, but also places themselves: places are always *in play*. In order to understand the tourism mobilities through which the contemporary Caribbean is constituted as a place to play, we must first understand how places are demobilized and remobilized, how they are 'put into play' with and against each other. The places of tourism are formed and transformed by changing infrastructures of physical and informational mobility, by cultural practices of travel and migration, and by the ever-shifting mobilities of people, markets, cultures, consumption, crime, risk, and disease. The making of places *to* play thus depends on particular ways of putting places *in* play.

Mapping Caribbean mobilities

Historically the New World 'plantation complex' (Curtin 1990) entailed a huge mobilization of plants, people, ships, material resources, foodstuffs, technologies, know-how, and venture capital. 'The sea became an economic highway for slavers, traders, buccaneers, and fishermen; then it became a passageway for escaped slaves, indentured labourers, and settlers; and later still, it was a watery flight path for emigrants and boat people' (Pattullo 2004: 339). The modern Caribbean, defined by its turquoise-blue sea and loosely tied together by shipping routes, airline networks, and radio, cable, and satellite infrastructures, came into being out of these mobilities.

Being more 'deeply and continuously affected' by migration than any other world region (Foner 1998: 47), it is said that the 'essence of Caribbean life has always been movement' and that 'movement is a significant and conscious reflection of identity in the Caribbean' (Duval 2002: 261, 263; Thomas-Hope 1992). Between 1950 and 1980, for example:

> about 4 million persons left the Caribbean to establish permanent residence else-where, principally in Europe and North America…[representing] 5 to 10 percent of the total population of nearly every Caribbean society, a higher proportion [of migrants] than for any other world area…. [In the mid-1970s] 16 percent of the total population of Jamaica and 25 percent of the population of Puerto Rico and Surinam, were living abroad.
>
> (Chaney 1987: 8–9)

Migration became 'a way of life and part of culture' in the Leeward Islands, according to Elizabeth Thomas-Hope, such that the 'concept of mobility…went deep into society's consciousness' (Thomas-Hope 1995: 174). And Huon Wardle points out a wide range of anthropological research into 'the structure of movement that is embedded in Caribbean life' (Philpott 1973; Thomas-Hope 1995; Foner 1978; Olwig 1993), suggesting that certain kinds of 'cosmopolitan subjectivities' have formed in the Caribbean, which involve 'an interweaving of physical movement and imaginative migration' through narratives of 'adventure' (Wardle 1999: 524–25). Many Caribbean novelists and writers maintain that to become truly Caribbean you must first go else-where; migration, exile, and return have become the grounds for forging a pan-Caribbean identity.

Caribbean mobilities were also crucial to the original theorization of transnation-ality, diaspora, and 'deterritorialization' (Glissant 1981; Gilroy 1993; Basch *et al.* 1994; Benítez Rojo 1996; Clifford 1997). The Caribbean has long been at the fore-front of cultural 'globalization', whether in terms of its uprooted people, Creole languages, or hybrid music travelling across world markets. Not only does each Caribbean society embody and encompass a rich mixture of genealogies, linguistic innovations, syncretistic religions, complex cuisine, and musical cultures, but these 'repeating islands' (Benítez Rojo 1996) have also exported their dynamic multicultures abroad, where they have recombined and generated new diasporic forms and places to play. It is argued, for example, that 'Caribbean life in New York City is the product of the continuous circular movements of people, cash, material goods, culture and lifestyles, and ideas to and from New York City and the islands and mainland territo-ries of the English- and Spanish-speaking Caribbean and, in recent times, the island of Haiti' (Chaney 1987: 3). New York can thus be considered 'the largest Caribbean city in the world' (Sutton 1987: 19). Caribbean experiences of transmigration contributed to the destabilization of nations as imagined homogeneous communities, and of migration as a linear movement from one nation-state to another (S. Hall 1990; Gilroy 1993).

Today the scripting and performance of the Caribbean as a glittering chain of beck-oning paradise isles again generates vast global mobilities of investment and disinvestment, tourism and migration, disease vectors, and environmental change. It is a dreamscape constituted by global movements of migrants, tourists, workers, entrepreneurs, and criminals; transnational flows of capital investment, monetary instruments, financial services, and laundered money; mediated flows of print and digital information, radio and electronic communication, intellectual property, and pirated products; and the unpredictable travels of global risks and threats to security (such as drugs, diseases, guns, and hurricanes). Tourism mobilities cannot be theorized in isolation from an appreciation of these diverse other mobilities, because the very places in which tourism comes to be performed are also places constituted by many other kinds of mobilizations and demobilizations. Many First-World mobilities are predicated on specific forms of demobilizing excluded places and populations – whether by dispossession, criminalization, or incarceration – and remobilizing these same places and populations for particular purposes (cf. Ahmed *et al.* 2003; Cresswell 2001). Indeed, the very 'place' we think of as 'the Caribbean' is as shifting as the sands, as wide as the oceans, and as liquid as the laundered money coursing through its off-shore bank accounts.

Global mobilities always take place (and make 'places') through located practices and material cultures – including those in 'peripheral' places. Tourism mobilities are not simply about the First-World travellers who tour the world, or the disembodied technologies and 'infostructures' that allow for their travel. Rather, as Kaplan argues,

> New technologies appear to promise ever-increasing degrees of disembodiment or detachment, yet they are as embedded in material relations as any other practices. They require hard industries as well as light ones...human hands to build the

machines in factories that are located in specific places regulated by particular
political and economic practices.... [H]ow do globalized information technologies, with their incumbent machinery and heavily freighted divisions of labour,
come to be characterized as transitory and light, as playful practices of subjectivity
that enable users to slip the moorings of location and materiality?

(Kaplan 2003: 207, 209)

The informatics industries that enable the conditions for instantaneous communication and seamless travel, including airline reservations, data processing, animation,
computer-aided design (CAD), telemarketing, financial services, business information
services, and database management, depend on low-wage, tax-free 'back-office
enclaves' such as the Jamaica Digiport International or the Barbados Data Processing
Centre to keep them flowing (Graham and Marvin 2001: 355–57; Skinner 1998;
Freeman 2000). Human hands (and bodies) are required to service tourism, to provide
hospitality and to perform the work that enables others to play. Like the hard industries that underwrite the ethereality of the information age, the labours of play are
embedded in material relations and located in specific places regulated by particular
political, economic, social, and cultural practices (Klak and Myers 1998).

We must consider not only the simple binary of mobilities and immobilities, but
also more complex dialectical process of *demobilizations* (actively engendered, often
resisted, and never fully effective) and *remobilizations* (including the pull of labour
markets and migrant diasporas, but also the constitutive discourses and practices that
enable the imagining, managing, and governance of mobility). As Appadurai notes,
there are 'increasingly complex relationships among money flows, political possibilities, and the availability of both un- and highly skilled labor...even an elementary
model of global political economy must take into account the deeply disjunctive relationships among human movement, technological flows, and financial transfers'
(1996: 34–35). Global mobilities of capital investment and disinvestment, including
illegal flows of drugs and money, and the global governance of such flows, have an
impact on the mobility of local currencies, inward investment, and labour in the
Caribbean, including its insertion into the high-tech service sector. The mobilization
of Barbados as a high-tech investment opportunity, for example, trades on its symbolic
capital ('little England') as a safe place for foreign investors, businesses, and visitors,
and as a playground for the rich and famous, from Madonna to Tony Blair. Its integration into the flows of finance capital, foreign investment, tourism, and information
associated with globalization, however, depends on a social, spatial, and cultural
disjuncture with the performance of labour by its 'high heels' pink-collar workforce
(Freeman 2000), whose spatial and social mobility is limited by the all-too-real
constraints of gender, race, class, and nationality.

The movement of money offshore from the major economies is ironically one of the
few comparative advantages available for small Caribbean states to exploit (Maingot
1993). As in the days of piracy, the Caribbean has come to be associated with illegal
movements of money into and out of its weakly regulated banking sectors, as well as
with forms of smuggling and drug-running which subvert (yet underpin) the formal
regulated economy. Several UK dependencies or former colonies are well-known tax
havens where unregulated banking greases the wheels of global financial velocity:
Anguilla, Belize, Bermuda, the British Virgin Islands, the Turks and Caicos, and

Grenada. With the current crackdown on the funding of terrorist networks it is becoming increasingly evident that 'liquid modernity' (Bauman 2000) requires complex checks and barriers to enable or constrain particular kinds of flows.

Tax havens such as the British Virgin Islands depend on 'massive flows of money into and out of their territory' (see Maurer 1995, 1997a, 1997b), yet as Maurer underlines, the key questions are: 'How do certain practices and processes constitute "capital" such that it can "move"? How do they also structure its "movements" so that they can have the sorts of effects that [the] globalization literature ascribes to them?' (2002: 103). Capturing highly mobile tourist expenditures within local economies, and the taxation of the hotel and cruise-ship industry, have become huge bones of contention in the Caribbean. For cash-strapped small island states cruise ships in particular 'are seen as moveable feasts that sail away into the sunset, their bars and casinos untaxed' (Pattullo 2004: 344). With 14.5 million cruise-ship passenger arrivals in 2000, the Caribbean has nearly half the world capacity of cruise 'bed days' (Pattullo 2004: 341), yet much of the profit they generate is ploughed back into non-Caribbean-owned companies, while island ports of call struggle to keep their 'product' (shops, services, facilities, and friendly people) up to international standards. Cruise lines can easily and suddenly cut destinations from their itineraries, or, even more troubling, can create their own 'places to play' by purchasing private uninhabited islands or anchoring off deserted stretches of beach to 'create their own version of paradise'. In many cases the ship itself becomes the destination and offers far more play-spaces (cinemas, ice-rinks, shops, bars, restaurants, sports facilities, and so on) than the islands themselves can provide (Pattullo 2004: 348–49).

The huge growth of the package-holiday and cruise-ship industries in the Caribbean has made it increasingly easy for tourists to get to the Caribbean and experience its charms in pre-packaged form. Forms of viewing nature are always ideological (Pratt 1992; Grove 1995; Poole 1998), and hence also encode relations of inequality between the viewer and the viewed in postcolonial contexts. Caribbean travel writing is deeply implicated in what might be called a politics of the picturesque, by which the framing of scenery became an exercise of colonial domination over Caribbean people. As Duncan argues, the picturesque 'is not simply a way of seeing, it is simultaneously a way of doing, a way of world-making' (Duncan 1999: 153). As a 'denatured' nature that has been 'renaturalized' for purposes of commodification and consumption (Franklin *et al.* 2000), the Caribbean is constantly reassembled as a primeval, untouched site of luxuriant profusion. This 'natural' assemblage is then used as the lure for economic 'development', military adventures, and tourist fantasies. Such renaturalized scenery then serves to stage the power relations between tourists and local inhabitants via tropes of 'natural hedonism', which script Caribbean bodies as sexual objects for the tourist gaze and as perpetual service-workers within the global tourist economy (Sheller 2004; see Kempadoo 1999; Mullings 1999). Tourists enjoy privileges of moving through the islands both by land and by sea, gaining a kind of overview that allows them to construct 'local' people as rooted to place, as apparently 'natural' as tropical nature itself.

The 'naturalization' of the social and economic inequalities of the contemporary tourist economy occurs through three steps: the objectification of Caribbean people as part of the natural landscape; the equation of that landscape (and hence those who people it) with sexuality and corruption; and finally, the marketing of the Caribbean via imagined geographies of tropical enticement and sexual availability. These fantasies of tropical nature are closely allied with objectifying practices of racial voyeurism and sexual tourism, which can be collectively referred to as 'hedonism'. The ability of the tourist to enjoy moving within and through the Caribbean depends on demobilizations of local people, who are barred access to resort areas except in so far as they perform service work: cooking, cleaning, providing entertainment, selling local crafts, offering local knowledge as guides and drivers, and selling bodies for sport and sex. The apparent freedom of movement and boundless travel in a 'world without frontiers' is thus produced by the techniques of binding people, places, and meanings in place, yet only in order to remobilize them in the performance of various services. Places to play are 'curated and fortified', in surroundings imagined as 'untouched yet within reach' (Titley 2000).

Another key aspect of the contemporary reproduction of these iconic Edenic islands is their transformation into spaces of irony, playfulness, and rule breaking. A playful orientalist tourism has been instituted in the West Indies founded on the global exoticism of cruise ships with names like *The Levant* and generic tropical resorts with 'Moorish' architecture and pan-Indian names such as *The Taj*. Such eastern fantasies are closely tied to ideas of luxury, sexual indulgence, excessive wealth, and corruption. Especially in celebrity havens such as Mustique there are numerous homes designed to project a fantasy Orient onto the Caribbean. The British theatre set designer Oliver Messel designed a series of villas there that 'are built along Japanese lines or in the Spirit of Bali, with names like *Moongate*, *Obsidian* and *Sleeping Dragon*'; David Bowie once owned one called *Mandalay*, while Mick Jagger has a 'Japanese-style' one. 'Balinese-style' villas in the Caribbean are taken to the ultimate on Richard Branson's $14,000-a-day Necker Island in the British Virgin Islands. As a recent travel writer typically observes in defence of such pleasure islands: 'The Caribbean is, after all, about indulging the senses: eating, sunbathing, sleeping and, of course, snorkelling. You don't exactly go there for the culture' (Pietrasik 2001: 12–13). Through this playful pastiche of material-semiotic practices the Caribbean is construed as mostly cultureless and thus all the more suitable for hedonistic 'natural' excess.

As Maurer argues of the British Virgin Islands (BVI), officials and business leaders market their territory to potential investors by inventing and making visible a British heritage for the tourist gaze, by 'crafting exclusionary citizenship and immigration laws' to deny certain people full political participation, and by 'writing laws, not to regulate finance, but to facilitate it' (Maurer 2002: 118). Following his analysis we could say that 'different sets of discourses around citizenship, finance, and nation congealed in the object' called the Caribbean, granting it 'temporary ontological stability and a certain temporal materiality' (Maurer 2002: 104). Particular Caribbean places to play 'congeal' at particular moments, and just as quickly may disappear. Thus Caribbean Paradise is performed through mobilizations of capital, demobilizations of labour, and remobilizations of colonial narratives, heritage, and built environments. Heritage takes strange shape here. A 2003 BVI Tourist Board advertisement, for example, appears like an old-fashioned map of the many small islands making up the nation, with a single red dot appearing on the largest island of Tortola (otherwise marked only with ridges and storybook names such as Shark Bay, Long Swamp, Smugglers Cove, and Monkey Point). The key and accompanying text indicate that the red dot represents 'the only reminder of what you come to forget. A single, solitary set of traffic lights'; there is no yellow diamond (multi-story parking); no blue square (shopping arcade); no orange line (motorway). So the tourist is invited to escape from civilization, to escape from the globally dominant forms of automobility, and to enjoy 'Nature's Little Secrets'.[1]

Besides 'untouched' nature and 'unspoilt' beaches (see chapter 3), histories of tropical decadence and pleasure are also redeployed in reinventing places to play. In the 1950s, for example, Havana was known as the 'brothel of the Caribbean' (Pattullo 1996: 90) and had up to 10,000 sex workers (Schwartz 1997: 122). This history contributed to the resurrection of sex tourism in Cuba in the 1990s (Rundle 2001), as Fidel Castro reopened the country to the international tourist market, and sold Havana as a destination trading on its crumbling colonial past, 1950s American cars (see figure 2.1), and the promise of sex. The sexualization of 'exotic' bodies has become a standard tool of Caribbean tourist promotion, and feeds into the development of sex tourism in the region (Clift and Carter 1998; Kempadoo 1999). Thus a sexuality drawing on the colonial past and its racial and gender hierarchies is coded into the representation of Caribbean landscapes and the tourist pleasures to be found there. Indeed, sexualized strategies for entering the extremely competitive international market for tourism are promoted by many Caribbean states, with suggestive names such as the St Lucian restaurant called Bang Between the Pitons, or marketing slogans such as 'It's better in the Bahamas'. As Jacqui Alexander argues, the Bahamian 'state actively socializes loyal heterosexual citizens into tourism, its primary strategy of economic modernization[,] by sexualizing them and positioning them as commodities'

Figure 2.1   Cuban cars

(Alexander 1997: 90). The state's orchestration of the pleasurable experience of being in the Bahamas reinforces the sexual hedonism of tourism by making it appear a 'natural' part of Caribbean culture. And as Sánchez Taylor has shown, sex tourism packages Caribbean people as 'embodied commodities' by turning the long history of sexual exploitation under colonial rule into a 'lived colonial fantasy' now available for the mass tourist consumer. 'A key component of sex tourism', she argues, 'is the objectification of a sexualized, racialized "Other", available at a low price' (Sánchez Taylor 1999: 42).

Demonstrating this connection between Edenism and hedonism, an article in a British Airways magazine concludes that in Jamaica 'you'll be in the nearest thing we have on earth to the Garden of Eden, and to make it even better, it's after Eve tempted Adam with the apple' (quoted in Kennaway 2000: 114). Such playful, light-hearted images can have very hard-felt consequences. Jamaica has a dual symbolic standing as a 'pleasure island' of laid-back escapism and the dangerous terrain of drugs dealers and political warfare (Aitkenhead 2000). 'The first time I went to Jamaica, I didn't know much about the place beyond a vague impression of pirates, palm trees, Noël Coward, ganja and beneath that a sense of intensity, a lurking voluptuous danger' (Jenkins 2001: 6). This frisson of danger adds allure to the place, trading on the extreme pleasures of resorts such as the infamous Hedonism II, wild-eyed ganja-smoking Rastas, and media images of 'Dancehall Queens' gyrating to erotically charged lyrics. Partly because of the mobilization of such images, however, Jamaicans are subjected to very high levels of international surveillance and control. The U.S. 'War on Drugs' stemmed from a panic over crack cocaine and Jamaican 'Yardie' gangs who were supposedly exporting violence to Britain and the United States (Joyce 1998; Gunst 1995). These dangers are then interpreted as a 'natural' feature of Jamaica's landscape and people, packaged as a hedonistic place to play.

Symbolically the Caribbean as a whole acts at once as a place of promising possibility (whether for profit or for pleasure) and as a risky and 'dangerous crossroads' (the trope of cannibals, pirates, and gunmen). Exclusions of categories of people from the otherwise easy flows of transnational tourism and elite nomadism are often linked to discourses of disease, whether it be the myth of Haitians as AIDS-carriers (Farmer 1992) or more subtle notions of cultural contagion and 'infectious rhythm' (Browning 1998). The global risks associated with criminal activities, gun crime, terrorism, environmental disasters, and other security issues are producing new modes of surveillance and governance of local mobilities within and out of the Caribbean region. Death always stalks those who travel in the pursuit of pleasure (see chapter 18).

Conclusion

I conclude, then, by drawing out certain precepts and premises for the study of tourism mobilities. First, mobilities are produced via complex assemblages. Mobility cannot be thought of apart from the infrastructures that support it, and these often involve certain kinds of demobilizations and remobilizations. Labour markets and immigration policies govern certain kinds of immobility or, more precisely, delimited mobility – because the 'local' populace are never entirely immobilized, and always have their own routes, migrations, and infoscapes. Places to play are always also places to work. Places to travel to are always also places of habitation. And both play and travel generate patterns of labour migration and transnational dwelling that must be understood together with tourism mobilities.

Secondly, this chapter has shown that the forms of detachment or 'deterritorialization' that are associated with 'liquid modernity' and its 'light capitalism' (Bauman 2000) are always accompanied by attachments and reterritorializations of various kinds. When bodies move they carry with them markers of gender, race, class, ethnicity, and nationality. When people move (whether as migrants, workers, or tourists) they also carry with them physical baggage, imaginative maps, and cultural schemas. Tourism is an embodied performance in which mobility is not liberation or escape from more fixed and stable identities, but rather reinforces and reproduces durable inequalities by reiterating and deepening disjunctures. Most importantly, perhaps, tourism is deeply situated in contexts of postcolonial legacies not only in the Caribbean but also in many other parts of the world. Places to play are continually reinvented, respatialized, and remobilized through the structuring narratives of colonial histories, through the legal frameworks of postcolonial law, and through the international frameworks of neo-colonial governance.

Whatever the extent to which the metropolitan imagination tries to mobilize the Caribbean in its own fantasy images, Caribbean cultures nevertheless escape and are remobilized in new and unexpected ways, which again feed back into the imaginary scapes of tourism. Diverse Caribbean cultural identities and artefacts travel the world, whether via the media, the internet and the world wide web, or packed away in the suitcases of informal commercial importers, music pirates, and underground transnational hucksters of 'local' goods such as callaloo and ackee. Caribbean musical forms repeatedly 'infect' the mainstream with their styles and sounds (Browning 1998). Caribbean creoles infiltrate urban spoken vernaculars through hybrid youth cultures at the street level. Caribbean literatures work their way into university syllabi and university English departments.

Finally, we can say that mobility is always located and materialized. Rather than the binary distinction between 'global mobiles' and 'local immobiles', I have tried to demonstrate that global tourism mobilities take place through mobilizations of locality and rearrangements of the materiality of places. Places to play are only contingently stabilized for purposes of touristic consumption, and these stabilizations require a wide range of other demobilizations and remobilizations: of labour and capital, of migrants and citizens, of images and information, of physical stuff and virtual realities. That is why we can think of places to play as places in play, always fluid, relational, and unstable.

Note

1   'The British Virgin Islands, Nature's Little Secrets', *Observer Magazine*, 30 November 2003, p. 58. See also www.bvitouristboard.com.

Chapter 3

Islands in the sun: Cyprus[1]

Richard Sharpley

Cyprus is pretty special...the fabled island of love is a rare jewel...this peaceful, stress-free oasis has...a long tradition of hospitality towards travellers that makes it one of the most welcoming holiday destinations on earth.

(CTO 2003)

Ayia Napa rocks 24/7 so if you want a peaceful holiday then book a coach tour with your Granny!...Both day and night the action never stops, the question is have you got what it takes?

(Club 18–30 2003)

Ayia Napa...used to be pretty much like every other resort in the Mediterranean. But that was before. Before the hype began.

(Cadwalladr 2001)

Introduction: islands as places to stay

Islands, according to King (1993: 14), are a 'most enticing form of land. Symbol of the eternal contest between land and water...islands suggest mystery and adventure; they inspire and exalt'. They have also long been popular tourist destinations. As Conlin and Baum (1995: 4) observe, 'the allure of islands...as places where people go for relaxation and rejuvenation has a long tradition which continues unabated'. The Romans, for example, built holiday villas on the island of Capri, whilst the birth of the modern phenomenon of tourism saw places such as the Isle of Man attracting significant numbers of tourists by the mid-nineteenth century (Lockhart 1997).

In more recent times, the development of international air transport systems combined with the emergence of an efficient and highly integrated tour-operating sector has enabled tourists to travel to ever-more-distant or exotic island destinations in ever-increasing numbers. Thus, since Vladimir Raitz's pioneering charter flight to Corsica in 1950 (Yale 1995), not only have Mediterranean islands (such as the Balearics, Malta, Crete, Corfu, Rhodes, and Cyprus) become some of the most popular short-haul holiday destinations for Europeans – and similarly, the Caribbean islands for North Americans and Indonesia (especially Bali) and the Melanesian islands for Australians and New Zealanders – but many have also become long-haul destinations. Examples of the latter include the Caribbean islands (for Europeans), the islands of the Indian Ocean (Mauritius, the Seychelles, the Maldives), South-East Asian islands such as the Philippines, Bali/Lombok, and the islands of Thailand (Koh Samui, the Phi Phi islands – the smaller of which was the location for the movie *The Beach* – and Phuket), and other, newly emerging destinations, such as Cape Verde off the Senegalese coast, and the Andaman Islands in the Bay of Bengal.

The attractions of islands as tourist places to play are diverse. The perceived physical characteristics of the islands where this tourist growth has mostly occurred – unspoilt beaches, clean sea, and an attractive climate – highlighting the intimate relationship between the land and the sea that surrounds it, are undoubtedly a major appeal, as are natural and cultural attractions. However, more important, perhaps, are the distinctive yet less tangible aspects of 'islandness' – the sense of distance, exclusivity, separateness, insularity, tradition and, frequently, colonial links – and the subsequent potential for tourists to permit themselves to indulge in liminal or compensatory forms of play (Gottlieb 1982; Lett 1983). In other words, while mainland coastal resorts may be places on the margin (Shields 1991), the physical separateness of islands places them well beyond the margin. They are places to which tourists are often attracted by the perceived opportunities for hedonistic or unlicensed playfulness, a fantasy 'other' affording opportunities for activities or behaviour unconstrained by the customs and structures of everyday home life. It is not surprising, therefore, that it is principally, though not exclusively, island destinations that have come to be associated over time with excessive or liminal behaviour on the part of tourists. Cuba in the 1950s was, for example, considered by American visitors as the 'nearby underdeveloped sub-tropical playground where anything goes' (Hall 1992: 110; Kolland et al. 2000; also see chapter 2 above), while in the 1970s, the Seychelles were considered the 'islands of love'. More contemporary examples include the 'Hedonism' resorts in the Caribbean (see chapter 2), the Greek island of Mykonos famed for its straight and gay clubs and nudist beaches, and Ibiza.

Of course, not all islands are associated with liminal tourist experiences, for the relationship between islands, their development as tourist destinations, and the roles that tourists play is enormously complex and variable. That is, the scope, scale and nature of tourism development, and consequential manifestations of tourist behaviour, varies not only from one island to the next but also within islands. The Balearic island of Mallorca, for example, has been eponymous with mass summer-sun package tourism since the early 1970s, yet this is largely restricted to the main resorts of Palma Nova and Magaluf, while other places, most famously the village of Deià, have long attracted a more exclusive clientele. Nevertheless, a number of general observations can be made about islands as places to stay/play that provide the framework for this chapter.

First, not only are islands popular tourism places but also, given the continuing spread of the 'pleasure periphery', an increasing number of island states are turning to tourism as a vehicle of socio-economic independence. Second – and as a corollary of this – the challenges typically faced by island economies, such as a limited resource base, dependence on the agricultural sector, the lack of a manufacturing base, and so on, have resulted in many islands becoming economically dependent on tourism (Bastin 1984; Wilkinson 1989; Milne 1992). For example, for islands in the developing world, 'gross receipts from tourism are larger than all visible exports put together' (King 1993: 28), while the world's top twenty nations ranked according to the proportion of the contribution of tourism to national GDP are all islands (WTTC 2001a). Moreover, given the political and economic structure of international tourism

(Britton 1991), island tourism development reflects the centre–periphery model of development (Milne 1997).

Third, and of greatest relevance to this chapter, tourism in islands needs to be managed if it is to optimize the benefits to the local economy and communities – in contemporary parlance, to be developed sustainably (Sharpley and Telfer 2002). Indeed, much of the recent island tourism literature focuses upon this issue (for example, Briguglio *et al.* 1996; Apostolopoulos and Gayle 2000; Ioannides *et al.* 2001). In particular, islands trying to address this issue hope to attract visitors who, in turn, contribute appropriately to local development, so that there are 'appropriate' performances by both hosts and guests. However, although as physical entities islands are relatively static, as tourist places they are dynamic (as are other, non-island destinations – see chapter 14 on the mobilities of Rio de Janeiro). Certainly, their insularity does not render them less vulnerable to global political or economic events that influence tourist flows, while destination-specific crises (such as the Bali bombing of 2002, political upheaval in Fiji in the 1980s and 1990s, the Turkish invasion of Cyprus in 1974) inevitably impact upon tourism mobilities in the affected places. Equally, the nature of the consumption of island tourism experiences is highly mobile – with consequential implications for the survival of the place as a tourist destination – and may often be influenced by an external (and internal) network of political, cultural, and economic forces that compete with local governance of the tourism sector.

This is very much the case in Cyprus. Since the invasion of 1974 and the de facto partition of the island that followed, the southern part of the island (internationally recognized as the legitimate political entity) has not only evolved into a popular mass-tourism destination but also, in a sense, has been transformed from 'somewhere' to 'anywhere' in a process that, arguably, reached its zenith in 2000 and 2001. Then, to young British tourists the island became synonymous with the resort of Ayia Napa, the party town that became the temporary summer home of the British 'garage' clubs – hundreds of thousands of young tourists flocked there not because it was Cyprus, but because of the 'garage' scene.

The cosy, family holiday image portrayed in the BBC series *Sunburn*, based on the resort of Paphos (see figure 3.1) was lost in the frenzy and hype surrounding the ascendancy of Ayia Napa as the 'new Ibiza' (a label that some would dispute – see Carl's Garage Site 2003) and its portrayal in the UK's Channel 4 television series *Ayia Napa: Fantasy Island* (2000). Since then, the club scene has moved on, Ayia Napa (though still popular) has been labelled by one commentator as the Mediterranean's second 'notspot' after Faliraki (Rhodes), and overall tourist arrival figures are in decline. In 2002, Cyprus suffered a 10 per cent fall in tourist numbers from the previous year, while arrivals in the first half of 2003 were 12.6 per cent lower still on the same period in 2002 (*Cyprus Mail* 2003).

Importantly, the way Cyprus has evolved as a tourist destination – especially since the 1980s – has been in direct opposition to Cyprus's official tourism development plans and policies. Moreover, not only does the current tourism policy appear increasingly unachievable, but the longer-term health of the tourism sector is now in doubt – it is a destination that has lost sight of its destiny. Therefore, I explore the factors that have contributed to these changing fortunes for Cyprus's tourism sector and, in particular, the network of external and internal forces that have constructed tourist play on the island. Much of the following relates specifically to the resort of Ayia Napa.

Although other resorts (such as Paphos [see figure 3.1] and Limassol) remain popular with their traditional markets, it is Ayia Napa that experienced the most dramatic swings in terms of visitor numbers and characteristics, swings that have had a knock-on effect upon the island's tourism industry as a whole. For the purposes of this chapter, therefore, I consider Ayia Napa as being synonymous with Cyprus.

Cyprus: a tourist island in the sun

At first sight, tourism in Cyprus has been a spectacular success. In 1960, the year of independence, the island was unknown as a tourist destination. As a relatively under-developed eastern-Mediterranean island state that, throughout the 1950s, had witnessed the political and, subsequently, armed struggle for *enosis* (union with Greece) as well as increasing inter-communal conflict between the ethnic Greek and ethnic Turkish population – the birth of the complex and as yet unresolved 'Cyprus problem' (see Attalides 1979) – Cyprus offered little to draw potential tourists to play. Indeed just 25,000 tourist arrivals were recorded in 1960 and its tourism 'industry', comprising mainly small-scale, family-run businesses located in hill resorts, was virtually non-existent (Ioannides 1992). Interestingly, such small-scale 'agrotourism' is now considered by the Cyprus Tourism Organization (CTO) to be one of the essential ingredients of the development of quality, sustainable tourism, though its potential remains limited (Sharpley 2002).

Forty years later, the island appears firmly established as a major Mediterranean summer-sun tourist place, an achievement all the more remarkable given that, following the Turkish invasion of the northern part of the island in 1974, the Cypriot tourism

Figure 3.1   Paphos Harbour, Cyprus

industry was rebuilt from scratch (see Andronikou 1987; Ioannides 1992; Seekings 1997; Cope 2000; Sharpley 1998a, 2001a). Arrivals, which had exceeded 2 million for the first time in 1994, reached a record 2.7 million in 2001 (a major driver being the 'Ayia Napa factor'), while receipts from tourism amounted to CY£1.227 billion (about $2.087 billion in 2000 values). Indeed, tourism has driven development on the island; tourist receipts represent over 21 per cent of GDP and over 30 per cent of GDP if the indirect contribution is included (WTTC 2001b). Around one-quarter of the working population are dependent upon the tourism sector and, principally as a result of the development and growth of tourism, the island's overall economy has grown at an average of 10 per cent per annum since 1980 (Ayers 2000). Per capita income on the island is now the third highest of all Mediterranean countries and, according to other typical development indicators, Cyprus enjoys 'high human development'. Indeed, in 1999 the island ranked 25th on the UNDP's Human Development Index, above Portugal, Singapore, and Malta. Cyprus now compares favourably with other developed nations and it is clear that this is directly related to the growth of mass tourism.

Despite this, not only does the sector face an uncertain future (the growth in arrivals resulting from the play place of Ayia Napa has masked an underlying stagnation in visitors), but more importantly, the direction of tourism development has significantly diverged from the path set out within official tourism policies and plans. In other words, and as the next section reveals, the manner in which tourism is produced and performed in Cyprus over the last four decades has been highly fluid, demonstrating a pattern relatively unrestricted by local planning and policy making.

Tourism in Cyprus: policy vs practice

Since 1974, not only has tourism been identified in all Cypriot national development plans as a primary means of achieving economic growth, but explicit proposals with respect to the scope and nature of tourism have also been included in those plans (see Sharpley 2001a, 2003a). The initial objectives were, inevitably, to reactivate the economy following the Turkish invasion and the subsequent partition of the island, by reinstating Cyprus on the global stage through the provision of accommodation, facilities, and related infrastructure necessary for international tourism.

By the early 1980s, however, the redevelopment of tourism was, in a sense, too 'successful'. Rapid growth in arrivals, fuelled by an equally rapid yet uncontrolled expansion of the accommodation sector, was accompanied by a lack of associated infrastructural development and inadequate environmental planning and control. Accordingly, a number of measures were introduced to refocus the direction and scale of tourism development away from the burgeoning coastal resorts and to restrict further increases in arrivals. These included financial incentives to develop accommodation facilities in the hill resorts and to restrict coastal development to five-star accommodation, as well as a marketing policy that targeted higher-spending, 'quality' tourists from diverse markets, that promoted the island's cultural attractions, and that sought to attract special-interest and low-season tourists. In short, tourism planning during the 1980s sought to relieve the increasing pressure of tourism development along the coast and to increase the value, rather than the size, of tourism at the same time as addressing its temporal and spatial problems, thereby reversing the island's increasing dependence on the rapidly growing, mass summer-sun market.

In practice, these objectives were not met. Not only did arrivals increase by an annual average of 16 per cent during the 1980s, but coastal accommodation supply, particularly in the lower-grade self-catering/apartment sector, also grew rapidly. In fact, some 30 per cent of the total increase in bedspaces between 1985 and 1990 was accounted for by new apartment accommodation while, by 1990, almost 40 per cent of the island's total accommodation stock was in Ayia Napa and nearby Protaras. Nevertheless, subsequent tourism development plans and policies have continued to embrace the same objectives of attracting diverse, higher-spending markets, encouraging special-interest tourism away from the coast (hence the emphasis on agrotourism as mentioned earlier), reducing seasonality and encouraging independent, non-inclusive (non-package) tourism. For example, the tourism section of the 1989–93 economic development plan recognized that 'if the present course of development is continued, it will in the long run have serious adverse effects on the competitiveness of our tourist product in the international market' (CPC 1989: 156). (As suggested shortly, this warning has now been realized – that is, Cyprus as a destination has lost the exclusivity that it arguably enjoyed both in the pre-1974 development phase and subsequently up to the 1980s, and is now in danger of falling 'out of play' in global tourism.) The plan went on to propose a nine-point strategy for the controlled, balanced development of tourism.

Similarly, the most recent tourism policy for 2000–2010 (CTO 2001) focuses on the development of 'quality' tourism based upon: maximizing income from tourism through a value-volume strategy (i.e. relatively lower growth in arrivals, but targeting higher-spending visitors); reducing seasonality; repositioning Cyprus's image as a destination, with greater emphasis on experiences linked to the island's culture and environment; and, generally, marketing the island as 'a mosaic of nature and culture, a whole, magical world concentrated in a small, warm and hospitable island in the Mediterranean at the crossroads of three continents, between West and East, that offers a multidimensional qualitative tourist experience' (CTO 2001: 33; compare with the image of the Caribbean as discussed in chapter 2).

Again, recent tourism development in practice contradicts these policies. Fluctuating arrivals figures during the 1990s were followed by steep rises between 1999 and 2001, largely related to Ayia Napa's brief reign as the Mediterranean's clubbing capital, while arrivals for 2003 may well fall back to 1998 levels. The island also remains highly dependent on British tourism mobilities and, importantly, on overseas tour operators who carry almost two-thirds of all arrivals on inclusive packages. This power of the operators enables them to dictate not only tourism development but also the nature of tourism consumption. Tourism also remains stubbornly seasonal, with around 40 per cent of all tourists arriving during the peak summer quarter (July–September). Nowhere is this more marked than in Ayia Napa which, in effect, closes for the winter. Tour operators do not offer the resort over the low season, since the family and principally youth market defines its tourism profile. In comparison, Paphos affords a year-round experience for the family and, during the winter, the 'grey' market.

Patterns of accommodation development are also at odds with 'official' policies. Not only does tourist accommodation continue to be spatially concentrated in the coastal areas but also, although more recent growth has focused on the higher-graded hotel sector, the island's accommodation supply remains weighted towards mid-range hotels and self-catering accommodation located in coastal resorts, underpinning the position of Cyprus as a summer-sun, mass-market destination. At the same time, the over-supply of accommodation has enabled tour operators, able to command heavy discounts, to reinforce the mass-market appeal of the island. Thus, Cyprus is consumed and performed as a mass, summer-sun destination for the British, with an image dominated by Ayia Napa's continuing, though declining, reputation as a party town. As a result, Cyprus lost its exclusivity; as a destination it is falling out of play in global tourism. The question to be addressed, therefore, is what forces determined the mobility of Cyprus as a destination and, in particular, the transformations in the nature of tourists' play on the island?

Cyprus: from 'somewhere' to 'anywhere'

Cyprus's position within the global tourism system has been, as we have seen, highly (and, most recently, downwardly) mobile. More specifically, Cyprus's image and consumption as a mass, summer-sun destination and, latterly, its brief popularity/ notoriety as a clubber's paradise or 'fantasy island' have developed largely in opposition to that envisaged by the island's tourism authorities in their policies and promotions. Certainly, the images portrayed in the TV series *Ayia Napa: Fantasy Island* (and the brochures of tour operators catering to the youth market) bear no resemblance to the distinctive Cyprus described in the CTO's own promotional material. As a result, the island has, to an extent, become an 'anywhere' beach destination, while the Ayia Napa club scene could, literally, be anywhere – the youth-tourist club market is the most mobile of all.

Two principal influences have contributed to this mobility of Cyprus as a destination. First, a network of forces inherent within the political economy of tourism production on the island all but eliminated the CTO's ability to govern the nature of tourism development. Second, this facilitated the evolution of certain tourist performances, in particular the Ayia Napa club scene, that transformed the island as a whole, and its consequent fragility as a tourist place. I focus primarily on the latter, although it is useful to review the factors that contributed to the relative impotence of the tourism authorities in Cyprus (see also Sharpley 2001b, 2003a).

The attraction of the island to British tourists is clear; it is familiar and safe, English is widely spoken, cars drive on the left, and the currency, the Cyprus Pound, has a familiar name. Moreover, cultural and historical links between the two states (dating from Cyprus's time as a British crown colony from 1925 to 1960, maintained partly by the continuing presence of two British sovereign military bases on the island) further contribute to a strong touristic relationship manifested in the supply of tourism services and experiences on the island.

More importantly, the power of the tour operators enables them to dictate the nature, scale, and seasonality of tourist flows to Cyprus, while an over-abundance of accommodation facilities resulted in downward pressure on prices and, hence, the continuing promotion of the island as a mass summer-sun destination. Thus, the ability of the CTO to reposition the island remains dependent upon the cooperation of tour operators in general and the manner in which they sell the Cyprus tourism experience in particular. Certainly, research has demonstrated that both mainstream and specialist UK tour operators typically promote Cyprus as a sun-sea-sand destination with an emphasis on fun, nightlife, relaxation, and a hint of romance (Sharpley

1998b), reinforcing the perception of island destinations as liminal, playful places – as having a quality of 'anywhere-ness'.

The ability of the CTO to direct tourism development in Cyprus is also hindered by local political structures and socio-economic tradition. With regard to political structures, Cyprus has a multi-tier democratic system of government, with a significant degree of authority delegated to the local level. There are, however, certain inherent weaknesses in the system. First, formal structures for the implementation of policy at the national level do not exist. Instead, there is a reliance on informal contact and agreement between political and industry leaders as opposed to a formal consultation machinery. This allows for political deals or favours that contradict or circumvent official policy, while also permitting conflicts of interest. Second, the multi-tier governmental structure, requiring elections at some level virtually every year, encourages policies and decision-making based upon short-term political motives at the same time as placing significant power in the hands of local municipal administrators, particularly mayors. Thus, the responsibility for planning decisions, infrastructural development, and other tourism-related activities lies with local politicians who, for electoral or other reasons, may not always make decisions in the wider regional or national interest.

The political structure in Cyprus is also suited to local social structures and traditions on the island, in particular the importance of land ownership within Cypriot society. To Cypriots, land is valued both as a status symbol and for its potential economic return, and many Cypriots wish to pass on their land to their children, as no inheritance tax is payable. Thus, the sale of land for hotel development has provided opportunities for rapid financial gain, or to buy more land as an investment. As a result, much of the coastal area of Cyprus has been developed for tourism relatively unhindered by planning controls, leading to the oversupply of accommodation noted above.

These political and economic influences that shaped the development of Cyprus as a destination in general are particularly relevant to Ayia Napa. The post-1974 drive to rebuild the tourism industry transformed the south-east corner of the island from a remote, undeveloped agricultural region into a purpose-built and rapidly growing beach destination. The lack of formal planning and control permitted the rapid, haphazard development of the area into a resort catering to the mass sun-sea-sand market for families and, specifically, younger tourists seeking nightlife experiences. Indeed, by the early 1990s, Ayia Napa was established as a major Mediterranean place in the sun as well as the principal centre for nightlife in Cyprus, though its reputation was not enhanced by occasional publicity surrounding the off-duty behaviour of British soldiers from the nearby army base. It has also grown into the island's largest tourist centre, accounting for more than a quarter of total tourist arrivals, though it remains seasonally defined – as previously noted, tour operators do not sell Ayia Napa holidays in the winter months. As a result, the overall success of Cyprus as a destination is disproportionately dependent on the success of Ayia Napa; recent falls in tourist arrivals in Cyprus are almost entirely due to a significant decrease in this one resort's popularity, although other factors, such as the relatively high cost of holidays in Cyprus and political instability in the Middle East, affect the island as a whole.

In short, over a twenty-year period, Ayia Napa enjoyed startling and mainly locally induced growth into a large and 'typical' beach resort consumed by those seeking fun, nightlife, and play while, interestingly, any negative socio-cultural impacts resulting from its development, at least up until the mid-1990s, were seen by many local people as outweighed by its economic contribution (Akis *et al.* 1996). The question that remains is why Ayia Napa recently experienced temporary status as a premier clubbing resort, only for it to lose that status, and along with it, much of its traditional tourism business as well? Ayia Napa had traditionally afforded experiences based upon typical beachlife and nightlife. But from the late 1990s, it became renowned as a nightscape (see Chatterton and Hollands 2003), its dominant attraction being the garage club scene, while the role of the beach itself was transformed. As Cadwalladr (2001) comments, 'this [Nissi Beach] is a beach unlike others in the Mediterranean. There are no families, no children...It's much cooler [than Ibiza]. Much more glamorous. Much more Dolce & Gabbana and Moet & Chandon and Versace and looking good.' Thus, Ayia Napa was, for a brief period, a 'cool' place (Skelton and Valentine 1998; also chapter13 below) dominated by tourist performances, in the clubs by night and on the beach by day. However, well-publicized incidents of drug-related arrests, violence, and sexual activity in public places deterred other tourists – families, couples and older visitors with no interest in the club scene – from visiting Ayia Napa, and they continued to stay away even after the resort lost its 'cool' status after the 2001 season when the clubbers began moving on elsewhere.

A significant feature of Ayia Napa's 'rise and fall' within the international club scene is that, prior to the late 1990s, the town was already a renowned nightspot providing a wide variety of club-based experiences. As Sellars (1996) observes, the potential of the youth club scene had long been recognized by the tourism industry and, from the early 1990s, major UK tour operators had developed specialist packages to Ibiza, Ayia Napa, and elsewhere aimed specifically at the dance-music market. Such holidays (and the material employed to promote them) served to heighten the sensational, ritualized experience of clubbing; that is, 'dancing can provide a release from many of the accepted social norms and customs of the "civilised" social spaces of everyday life, such as social distance, conformity and reserve or disattention' (Malbon 1998: 271), reflecting precisely and magnifying the liminal experiences afforded by island tourism.

However, from 1999 to 2001, Ayia Napa enjoyed an enhanced reputation as the coolest club scene in the Mediterranean (garage music itself being regarded, at that time, as the coolest sector of the dance scene). The reason for this remains unclear. Certainly, the production of the branded, themed nightscape typical of many urban places (Chatterton and Hollands 2003) was not mirrored in the resort although it was, nevertheless, the 'selection' of Ayia Napa by the DJs from major London clubs that drew clubbers in their thousands. Consequently, Ayia Napa became a short-lived international brand, but a brand that the resort had little control over. In other words, the elevation of Ayia Napa's clubbing status was not planned or managed by the local tourism industry – it 'just happened' (Cadwalladr 2001).

The subsequent screening, in 2000, of the TV series *Ayia Napa: Fantasy Island* both heightened the resort's temporary popularity and hastened its downfall. It publicized not only the ritualized performances in the nightclubs but also, more significantly, the organized (by tour reps) 'fantasy' parties and events that promoted heavy drinking and sexual activity. It conveyed, therefore, contradictory messages – the reality of an established, 'cool' and ritualized dance scene that transferred itself to the island for the summer months combining with the hedonistic, unlicensed behaviour of young tourists on the fringes of the dance scene. As a result, Ayia Napa was heavily promoted (and still is) by tour operators seeking to attract the youth market – the Freestyle (2003) brochure, for example, headlines its Cyprus section with 'top djs and promoters have flocked to Ayia Napa and with all the gorgeous beaches, bodies and bars who can blame them?' The 'cool' clubbers have, however, moved on to the next cool place (wherever that may turn out to be), as have the resort's other more traditional visitors such as younger families and couples. As with Port Vell in Barcelona (see chapter 12), Ayia Napa has become 'uncool' and, consequently, the resort and the island face a very uncertain future.

Conclusion

To conclude, islands in warm climates have long been popular tourist destinations, their attraction being a function of both the tangible characteristics of climate and geography and the less tangible sense of 'islandness'. It is, perhaps, the latter that has underpinned the success of island destinations, their physical insularity, distance, and separateness translating into opportunities for liminal playfulness on the part of tourists. Nevertheless, the success of islands as tourism destinations is never assured; they are unstable places to stay/play, demonstrating a dynamism that is responsive to transformations in the performances of both hosts and guests. These may, in turn, be influenced by a network of forces both within and beyond the destination – in the case of Cyprus, ineffective governance of tourism has caused it to succumb to the external forces of tour operators and the restless international club scene. This means that the island is ultimately in danger of falling 'out of play' within international tourism. There is a need, therefore, for islands (and, indeed, all destinations) to anticipate and to respond flexibly to the constantly shifting network of forces, to ensure that appropriate performances on the part of the hosts meet those of tourists, and vice versa.

Note

1   Unless otherwise stated, this chapter is concerned with tourism in the southern (Greek-speaking) Republic of Cyprus, which is internationally recognized as the legal government of the island. At the time of writing the island remains divided, although accession to an enlarged European Union in 2004 may accelerate the creation of a reunified federal state.

Chapter 4

Ecotourists on the beach

Rosaleen Duffy

Introducing ecotourism

Ecotourism is one example of a new networked pattern of social life indicative of increasing global mobilities. Through global travel, and the extension of ecotourism, destinations in the global periphery are being transformed into global places to play. The ways that places in the South are redefined and recreated as global leisure sites are clear in how visitors from the North talk about their chosen destinations and the multiple roles that they perform while on vacation.[1] The global places to play examined here can be seen as discursively constituted through ecotourist talk, and recreated through tourists' performances and practices. It also highlights the multiple narratives of nature and culture contained within the ways ecotourists talk about their chosen destination and host culture (see Macnaghten and Urry 1998).

Since this chapter is concerned with ecotourism performances and practices, it is useful to have a working definition. Defining ecotourism is difficult, and as a result it has become a loose, catch-all term for tourism concerned with visiting and experiencing some aspect of the natural environment, be it wildlife, rainforests, coral reefs, or even beaches. Boo defines it broadly as nature tourism that consists of travelling to a relatively undisturbed or uncontaminated natural area with the specific object of studying, admiring, and enjoying the scenery and its wild plants and animals as well as the existing cultural manifestations in the area (1990: 1–3; also see Bottrill 1995: 45–54). Along similar lines the Ecotourism Society defines it as responsible travel to natural areas that conserves the environment and sustains the well-being of local people.[2] Furthermore, ecotourists themselves are a relatively difficult type of global traveller to categorize, and can include everyone from backpackers on a $10 per day budget to high-spending Galapagos cruise passengers. However, in general ecotourists can be described as vacationers with an interest in outdoor pursuits, who are financially comfortable, well-educated, older people with free time to travel (Ballantine and Eagles 1994: 210–12).

This chapter is concerned with how ecotourists construct and define their experiences of their chosen destination, and then links that into ecotourist behaviour and the ways they transform the environments they visit. The ways that ecotourists re-order space and place in the South is examined through the case study of Belize. Belize has marketed itself as an ecotourism destination, with its main attractions centred on terrestrial environment such as rainforests (on the mainland) and beach/marine attractions (on its many islands, or Cayes). This chapter focuses on ecotourists visiting the islands off the mainland coast of Belize, and the transformations they produce in those islands. In this chapter I first examine how ecotourists talked about their holiday experiences in terms of seeking independent travel, motivations for choosing a destination, and the importance of the local environment. Second, I provide an analysis of the environmental behaviours displayed by ecotourists and their effects on local environments and tour operators. Finally, I investigate the broader impacts of ecotourism as it transforms local environments into the ideal beachscapes demanded by the global tourism industry.

Ecotourists, the environment, and global travel

Despite the difficulties in defining ecotourism, it is clear that ecotourists are a relatively new and rapidly growing group in global travel. This is directly related to changes in social and economic structures in Northern industrialized societies. Mowforth and Munt suggest that the more familiar mass tourists' interest in the three 'S's (sea, sun, and sand) has been replaced by the independent tourists' enthusiasm for the three 'T's: trekking, trucking, and travelling (1998: 125–55). Ecotourism is attractive to particular social groups, and the demand for ecotourist holidays seems linked to the environmental movement, which constitutes a new radicalism among the middle classes of the western industrialized world. Urry suggests that contemporary tourist practices are cultural in that they comprise signs, images, texts, and discourses. In addition, he argues that identity is now formed through consumption and play rather than through employment (Urry 1994: 233–36). In this way a holiday constitutes a positional good where destinations are consumed because they are meant to convey superior status and taste (Urry 1990: 44–47).

The choices of destination and activities undertaken on vacation are presented as signifiers of the individual's merit as a traveller rather than as a 'vulgar' tourist (see Buzard 1993). This is also reflected in the switch from mass tourism to more tailor-made forms of independent travel ranging from escorted tours to cookery holidays (see Urry 1990: 47–50; Cotgrove and Duff 1980: 341–48; Munt 1994a: 101–105; and Beck 1995: 53–57). Munt (1994a) suggests that this indicates a desire for holidays in keeping with postmodern culture. For example, tourism practices are increasingly about travel in conjunction with other special-interest activities, such as scuba diving, cycling, or working on a conservation project. This means that tourism itself has become increasingly interwoven with education and learning new skills, self improvement, and enhancing a sense of self-identity – again in keeping with broader postmodern cultures. This is clear from the development of activity or educationally based tours such as Antarctic cruises complete with a staff of expedition leaders, naturalists, and lecturers who add an educational dimension to the tour.[3] Furthermore, the rise in conservation holidays can be explained as part of the increasing interest in independent vacationing that uses travel as a form of self-improvement. For example, in Britain there has been an expansion in interest in actively volunteering for charities and associated working holidays rather than donating money.[4]

This chapter is concerned with how the ecotourists I interviewed in Belize constructed their holiday experiences. Such travellers' tales show how ecotourists conceptualize and define their travel choices and their destinations, as well as how that is related back to global ideas/narratives of the ideal beachscape or holiday activity. Bauman (1995) argues that we live a 'Life in Fragments', where our self-identities are increasingly being pulled apart and remade by the pressures of postmodern existence. People inhabiting this postmodern world find themselves confronted with multiple options that allow people readily to exchange one identity for another (Zimmerman 2000: 123). It is important to note here that ecotourists are not a homogeneous group, performing a single identity. Rather, there are multiple types of ecotourists (interested in coral reefs, or rainforests, or adventure activities, and so on) who perform different and multiple roles at various times during their holiday (see Cohen 1979; Wickens 2002; McMinn and Cater 1998; Gibson and Yiannakis 2002; and Macnaghten and Urry 1998). Nevertheless, the ways that ecotourists conceptualize, define, and describe their vacation experiences reveal that travel stories assist individuals in narrating self-identity, those identities are then affirmed and contested through discussion of 'travellers' tales' among fellow tourists, to peer groups at home – or even to an academic interviewing them during their holiday (see Desforges 2000; Elsrud 2001; Murphy 2001; Galani-Moutafi 2000).

Since ecotourists are defined as being especially interested in visiting rare natural environments, yet reducing the impact of their holiday-making, it might be expected that reefs, landscapes, and wildlife would constitute a major motivating draw. Engagement with travellers and their tales in Belize indicated that there were a number of common and recurring themes. One was their motivation to travel to Belize, which ranged from a wish to experience a new culture to wanting a challenging and unusual holiday. Broadly, ecotourists had been attracted to Belize through friends or relatives who already had experience of the place. This was also reflected in ecotourists' interest in reading about a destination before they arrived, and they were generally informed about Belize by guidebooks, especially the relevant *Lonely Planet* and *Rough Guide*, which provide much more information than travel agents (interviews 1, 4, 12, 33).

However, some of those interviewed also remarked that they did not know a great deal about Belize beforehand, and had expected it to be more like the rest of Central America and less like the Caribbean (interviews 13, 17, 18, and Buzard 1993). A significant number stated that their primary motivation for coming to Belize was to dive on the coral reefs. Belize is marketed as a scuba-diving destination, and is well known among diving enthusiasts (interviews 5, 13, 20).

The second group were tourists (often, but not exclusively, backpackers) who were on a longer trip through Central America, and they mentioned that the attraction of Belize was as a Caribbean beach destination where they could have a break and relax during their travels through the region (interviews 1, 2, 14, 21, and see Elsrud 2001; Murphy 2001; and Hampton 1998). Some of them did dive while in Belize, but others were satisfied with snorkelling on the reef or river trips to see manatees and Mayan ruins (interviews 6, 33, 35).

In general, ecotourists had a list of must-see sights in Belize. One ecotourist stated that his trip was motivated by a desire to see a big cat: 'that was the reason for coming to Belize: I want to see a jaguar. It is that thing of ruins, rainforest, and reef' (interview 4). For many ecotourists it was clear that 'environmental' attractions constituted the core motivation for travelling to Belize. Being thought of as an environmentally aware traveller is an important signifier of social position and a commitment to environmental beliefs. For others an independent holiday was all about retracing the steps that they had travelled in their youth (interview 7). One couple stated that now they were both earning a salary they were willing to pay a little bit more to be independent (interview 15). Part of the attraction of travelling was that they could return to their home societies with a new set of travellers' tales about the great moments and total disasters they experienced in off-the-beaten-track locations (interview 16). In addition, ecotourists were keen to point out that they tended to buy a plane ticket first and decide where to stay once they had arrived. A number of those interviewed suggested that they liked to determine their own pace of travel and make their own decisions rather than being dictated to by a package-tour operator (interviews 3, 8, 9, 12, 21).

Ecotourists presented an image of themselves as people interested in the process of travel for its own intrinsic value, since their holiday experience was about self-improvement or even self-recreation. This clearly intersects with Edensor's (2001a) argument that this kind of holiday is connected to a 'can-do' philosophy where ecotourists lay claim to their superior status through physical achievement. Holidays then become part of a rite of passage, where ecotourists set themselves a series of 'over-comeable' challenges that test their adaptability, endurance, and initiative, but are ultimately designed to be achieved with a moderate (but largely comfortable) level of difficulty (also see chapter 7). The combination of holiday time with learning experiences or character-building activities was apparent in their activities. For example, many used their holiday in Belize to learn to scuba dive or to gain more diving experience (interviews 22, 23, 34). The idea that ecotourists will engage in self-denial to ensure that their vacationing has minimal negative environmental impact is evident in the forms of travel chosen. However, in choosing what is packaged and sold as independent and rugged trekking, ecotourists deny themselves the comforts of luxury travel as part of a journey of self-discovery and self-development. One interviewee stated that he had previously travelled to Southern Africa and liked this form of travel because it involved both discovery and self-discovery (interview 23).

The challenging experiences associated with more independent forms of travel were vital in determining whether ecotourists enjoyed their vacation. These challenges were clearly designed to be over-comeable with reasonable effort in the end, while providing the necessary holiday thrills and satisfying desires for personal development. One interviewee remarked that the problems he had faced while travelling, while unwelcome at the time, were what really made the experience enjoyable and memorable. His flight had been disrupted in Corozal, an event which he did not enjoy at the time, but he felt that 'once I am home I would not change all the experiences for the world. It is the experience and I would not miss it when I come to places like this' (interview 10). The importance of travellers' tales centred on travel disasters, and the individual's response to them can be related to the need to think about tourism as character-building experience and something to impress one's peers (also see Lee and Crompton 1992: 732–37).

This desire to use travel experiences for self-creation and re-creation has been subject to some critical analysis. For Munt the new middle-class tourists from western societies have increasingly aestheticized and fetishized developing countries as tourist destinations, and glossed over the realities of life in the South (1994a, 1994b: 49–60; and Mowforth and Munt 1998: 125–87). Munt argues that ecotourists are better defined as ego-tourists because they search for a style of travel that is reflective of their own perception of themselves as having an alternative lifestyle, and which is capable of enhancing and maintaining their own cultural capital (1994a: 105–108). Munt's arguments were clearly illustrated by some of the ecotourists interviewed, who were keen to distinguish themselves from mass or package tourists. For example, a number pointed out that they were not 'Cancun types' and they differentiated themselves on a social level through consumption of a holiday they perceived to be more 'challenging'. One interviewee indicated that the type of person attracted to package tours was quite different to herself since she identified package tourists as 'dodgy people' (interviews 5, 20, 24, 33, and see Munt 1994a: 101–23). These ecotourists viewed their leisure time as a matter of personal status, and held that consumption of a holiday in a particular destination, with carefully chosen activities, is one way in which people can project an image of themselves to their peer group. As Arellano points out (chapter 7), the very public exhibition of endurance turns touring into performing, thereby transforming the metropolitan body into a moving corporeal body.

Furthermore, fetishization of developing countries assists in constructing and perpetuating an image of a destination as a premodern society inhabiting an idealized tropical beachscape and primordial rainforests. This indicated a desire to view and experience an ecotourist beach utopia. As Sheller notes (chapter 2) the discursive formation of Caribbean scenery is related to ideas of hedonism, where the destination represents relaxation, rejuvenation, and sensuous abandon. The landscape and culture are repeatedly represented as Edenic, or a heaven on earth, and ecotourists in Belize regularly talked about their chosen destination in these terms. Some stated that Belize had conformed to their expectations of the Caribbean as constituted by islands and sunshine (interviews 1, 4, 24, 34). One remarked that 'my image was pirates, seriously, and palm trees' (interview 16). Another part of the image is of the Caribbean as filled with reggae music and as a 'relaxed place to chill out' (interviews 10, 16, 33), or as being 'unhurried, where tourists are forced to relax' (interview 19). Ecotourist expectations, and the imagery they used to describe the islands, reflected the stereotyped representations of the Caribbean as a destination with silver sand, palm trees, and turquoise water.

Ecotourists thus indicated that they continually made and remade their identities through performing different roles (independent travellers, rugged trekkers, scuba divers, environmentally conscious or culturally aware travellers, and so on). In this way they conformed to the argument that postmodern society allows people readily to exchange one identity for another. However, the performance of environmental consciousness and self-limiting behaviour does have a role to play in how the environment is being transformed by global tourism. Rather than creating an environmentally benign form of tourism, ecotourism and ecotourists have multiple impacts on the environment that are the result of a complex interaction between ecotourist demands and the global tourism industry. How ecotourists perform their identities in relation to a place has transformative effect in the South.

The environmental impact of ecotourists

The idealized views of ecotourists have a direct impact upon the environment in Belize. The discursive constitution of nature as Eden or a tropical paradise (cf. chapter 2), plus ecotourist performances and practices, play key roles in transforming local environments into places to play. It was clear from the interviews that ecotourists were very interested in enjoyment, relaxing, diving, or snorkelling, and they were not greatly concerned with the negative impacts of tourism. Instead they glossed over the apparent problems and realities of a local development policy. In this way they mirrored the attitudes and impacts of the mass-tourism counterparts they often disparaged (Buzard 1993).

Ecotourists possessed some environmental knowledge about Belize, but this did not extend much further than an awareness of the fragility of the coral reefs. Again, this was related to the idea that tourists were having a 'getting-away-from-it-all' experience, since ecotourists expressed a desire to forget about environmental problems in their pursuit of enjoyment. Ecotourist knowledge about the coral reef came from two main sources: guidebooks or background reading; and from tour guides once they were in Belize (interviews 5, 21, 24, 33). The majority of ecotourists knew that a trip to Belize meant being able to snorkel or dive on the second-largest barrier reef in the world (interviews 25, 26). However, experienced divers demonstrated a wider knowledge of reef ecology and of the activities that could not be allowed on the reefs. One interviewee was concerned at the amount of damage caused by learner and newly qualified divers because of their inability to control their buoyancy; it was common for such divers to crash into coral, or hang on to it to halt an untimely ascent or descent (interviews 11, 17, 27, 28).

Most ecotourists stated that it was important for individuals to behave in an environmentally conscious way, and in many ways they perceived it as their contribution to conservation in Belize. This idea of self-limiting consumption appeared to be popular with ecotourists (interviews 6, 12, 26, 19, 21), yet despite this it was also argued that individual efforts, while necessary for marine conservation, were not sufficient on their own, so government and private-sector regulation were also necessary. One couple remarked that if ecotourists were ignorant of the environmental sensitivity of the reef, then it was up to the tour guides and government to provide information and control ecotourist behaviour on the reef (interviews 2, 6, 16, 29). Ecotourists were concerned that the main attraction – the coral reef and marine life that it sustained – were conserved. In many ways this concern was motivated by a desire to ensure that the reef was still a beautiful object to see on subsequent holidays rather than conservation for its intrinsic value (interviews 30, 33). Ecotourists also expressed their motivation for coming to Belize to see the reef before it was destroyed or before it became too overcrowded with other tourists (interview 31). Here again, reef conservation was clearly constructed and justified in terms of group satisfaction, since saving the reef was important so that future generations could visit it.

Although it might be expected that ecotourists would display a higher level of environmental awareness and concern, the demands made by these visitors to Belize played a key role in transforming the local environment. Ecotourists colluded with ecotourism operators to create a staged authenticity so that visitors enjoyed their vacation (on authenticity, see Edensor 2000; Wang 1999; and MacCannell 1989). For example, it was thought desirable that tour operators provide visitors with a list of must-sees and must-dos. When diving or snorkelling ecotourists were keen to see big, charismatic marine life such as sharks or deep-sea barracuda (interviews 22, 27, 43). Those involved in ecotourism and conservation in Belize have recognized the attractiveness of allowing ecotourists easy access to spectacular or iconic marine creatures. The importance of Shark Ray Alley and Hol Chan Marine Reserve trips for snorkellers, and the Amigos Wreck dive or Blue Hole dive for divers, illustrates this. Shark Ray Alley and the Amigos Wreck are well known in San Pedro town (Ambergris Caye) for the relative ease of shark and ray spotting. In addition, the Amigos Wreck was specifi-cally sunk by the Amigos Del Mar Dive Shop because ecotourists enquired about wreck dives in the vicinity of the island. It was also intended to reduce pressure on other popular dive sites that were becoming overcrowded; currently the Amigos Wreck is one of the most frequently requested ecotourist dives (interview 44).

When the weather was unsuitable for diving and snorkelling there was a feeling of frustration and a pent-up demand for diving. As a result, ecotourists requested dive shops to run a tour even during very poor weather because they had specifically come to dive each day (interviews 27, 33). In response to this, tour operators and conserva-tion groups were involved in a plan to create an artificial reef to attract marine life on the sheltered mangrove side of Ambergris Caye, because it would allow ecotourists to dive and snorkel regardless of weather and sea conditions (interview 43). These exam-ples show how ecotourist performances and practices transform local environments into their ideal holiday paradise. The demand for minor over-comeable challenges (such as moderately difficult wreck diving or feeding rays) and a bodily experience of a particular place has transformed the environment of Ambergris Caye as businesses responded to demand. The local marine environment has been modified to satisfy ecotourist desires.

One key debate in the ecotourism industry centred around touching the environ-ment. The issue of touching marine life was one that concerned tour guides and divemasters. Some guides remarked that although ecotourists were told not to touch the fish they still went ahead and did so (interviews 35, 41). When interviewed, ecotourists made it clear that they were not in favour of touching rays, sharks, corals, or any other forms of marine life. However, from interviews with tour guides and from observation it was clear that ecotourists over-reported their adherence to environ-mental behaviour (see Scott and Willits 1994: 248–50). However, some tour guides did not inform their clients that they should not touch the coral or the fish. In fact tour guides that were more environmentally aware were also concerned that unless other guides were more vigilant then the basis of their livelihood could be destroyed (interviews 35, 41). Again this highlights the desire to transform the metropolitan body into a corporeal body, capable of sensing the environment through experiencing, touching, smelling, or tasting the 'other'.

However, whether licensed and trained or not, all guides face complex problems in dealing with ecotourists. This is partly because they are not a captive audience, which means that the guides have to make the trips entertaining and interesting. Yet guides also have to tackle difficult issues, such as ensuring that visitors are aware of the correct behaviour towards the environment and especially wildlife. At meetings of the Tour Guides Association in San Pedro, the issue of the correct briefings for visitors taking trips to the coral reefs constantly arose. One guide suggested that there were always arguments about the guides and tour shops that simply allowed ecotourists into the water with no environmental briefing and no support if they got into difficul-ties and had to hang on to coral heads. However, among guides that did provide an environmental briefing and asked the ecotourists not to touch the reef, there was a concern that ecotourists simply did not heed their advice, even when told that touching the coral would kill it. Tour guide Daniel Nunez stated that he gave an envi-ronmental briefing on the boats before snorkellers entered the water. Ecotourists were asked not to touch or stand on the coral and to signal for assistance rather than hang on to corals when tired. However, despite a broad knowledge of the coral reef and an interest in local ecology, the ecotourists did not heed this advice – there was an element of 'I have paid for this so I will do it' (interviews 35, 40, 41, 45). These debates about how ecotourists behave, and the responses of guides, demonstrate the power ecotourists have in a local context. Their economic power has led to significant transformations of place and space, and tour guides as well as local conservation organizations can often be disempowered in the face of the purchasing power of 'environmentally sensitive travellers'.

One scuba-diving and snorkelling guide remarked that he thought he had seen a lot of environmental change at Shark Ray Alley. The site is popular among ecotourists because it is where nurse sharks and stingrays congregate to be fed. Originally the sharks and stingrays had gathered there because it was where local people gutted and cleaned their catch. As ecotourism developed in the area, the site was presented as an opportunity to swim with, touch, or hand-feed the sharks and rays. Diving guide Oscar Cruz suggested that the sheer number of visitors stirring up the sandy bottom as they swam, snorkelled, and dived caused pressure on the site because the sand smothered the coral heads (interviews 39, 46).

In addition, one tour guide suggested that as soon as ecotourists saw manatees and especially dolphins in the water, they wanted to jump in to experience swimming with them. However, this had provoked a debate about reintroducing trips marketed as a 'swim with the manatees' experience.[5] Currently, the manatee is a protected species and the government of Belize does not allow tour operators to run trips that involve swimming with them. Instead, ecotourists are allowed to view manatees in a boat with its engine turned off, so the animals can only be approached by using a pole to move through the eelgrass beds and mangrove swamps (interviews 37, 38). Initially, it was the guides in San Pedro and Caye Caulker that decided to stop swimming with manatees, without any prompting from the conservation bodies, after discovering that the manatees had left those areas under most pressure from ecotourists. Consequently the guides lobbied the Government and Coastal Zone Management Plan to prevent swimming with manatees (interviews 35, 37).

The debates over manatee trips indicate that host societies have to deal with a number of complex interactions between ecotourists and the environment. The difficulties raised by the ecotourist desire to get physically close to – and to touch – marine wildlife demonstrates that it brings peculiar problems. In particular, tour operators, needing to generate a profit, felt under pressure to allow visitors to get physically close to marine life even if it has a negative environmental impact. In addition, some guides indicated their disapproval of swim-with-the-manatees trips because they are also interested in protecting their business's financial viability over the longer term.

Beyond transforming and changing the trips on offer, ecotourism has a more indirect impact. It transforms environments that were previously not part of the tourist itinerary into places to play, and in so doing landscapes and environment are restyled to conform to the idealized images. These impacts may be driven by the demands of Northern holidaymakers but are shaped through processes of globalization, especially the interlinkage of global real-estate developers and local business people. These processes are clearly beyond the control of ecotourists at an individual level. Even with ideas of green consumer behaviour, most visitors to Belize do not have the time, or perhaps the interest, to choose the locally owned hotel, to check if the beach has been artificially created, or how sewage is disposed of. Even if ecotourists are highly concerned with reducing the environmental impact of their holiday choices, they can only choose from a range of options provided by tourism businesses either in their home societies or the visited countries.

McCalla suggests that while developers make adequate provision for guest welfare, the common failure is weak environmental management during the phase of hotel operation. For example, dredging to provide ecotourists with sandy beaches, and the dumping of construction waste, destroys aquatic habitats and wildlife. In addition, the replacement of mangroves with concrete walls and coastal roads destroys coastal ecosystems that rely on a symbiotic relationship between mangrove swamps and coral reefs. Equally, the conversion or drainage of coastal wetlands into building land has radically changed the aquatic ecosystem (McCalla 1995: 61–63; and Maxwell Stamp Plc 1991: 19–26). The ecotourists interviewed were unaware of this type of indirect effect of their presence because when they arrive in Belize the services are already in place.

This reconfiguration of the environment for ecotourism had also led to an illicit mobility of sand. The increasing phenomenon of sand pirates in Belize had raised concern among local conservationists. The sand pirates operated at night and removed sand from neighbouring islands, which was then used to build up artificial, but aesthetically pleasing, beaches on coral atolls that were naturally sand-free.[6] This phenomenon has arisen because of the ecotourist desire to have destinations that conform to their stereotyped image of a pristine Caribbean paradise of turquoise water, white sand, and coconut palms (or to view it from a different angle, because of local developers' desire to provide destinations that conform to the stereotyped image that they believe tourists are seeking). Ambergris Caye has suffered from the activities of sand pirates more than other islands. The sand is not a renewable resource in a conventional sense (that is, on a human timescale), but is an important part of the wider ecosystem on the island that supports coral reefs, mangroves, and a variety of fauna, and so sand piracy has a significant negative impact on the environment. The global demands from ecotourists for their own definition of a tropical paradise, complete with the idealized 'global beach', has led to local processes that significantly transform environments. These new artificially created beach utopias are represented by the global and local tourism industries as natural, authentic, and unspoilt.

The environment in Belize has already been transformed by ecotourist use. This is in one sense inevitable because the environment would always have to be modified to some extent in order to build accommodation and serve food to visitors. However, the interesting issue is how ecotourists still view such environmental changes as part of the natural ecology of their chosen destination. Despite the obvious remodelling of the environment to suit ecotourist needs, such as removing eelgrass to create perfect silver Caribbean beaches, visitors to Belize still refer to it as unspoilt and underdeveloped (interviews 1, 3, 20). This is partially because Belize promotes itself as an unspoilt and pristine ecotourist destination, and partly due to the obvious contrasts with the forms of development in the industrialized countries that ecotourists hail from. Ecotourists are a major cause of ecological transformation in Belize, yet they are not able to exercise control over tourism developments.

Conclusion

Ecotourism is one example of a new globalized and networked pattern of a mobile life. Ecotourists discursively construct the ideal tropical paradise, but in so doing they have a physical impact on the local environment. In this way Northern holiday tastes and demands transform space and place in the South, recreating its landscapes into a globally approved idealized beachscape. Examining the case of Belize demonstrates that ecotourism and ecotourists transform local environments into global places to play. Through increased global mobilities and the associated expansion of long-haul travel, places in the South have been reformulated and relandscaped to create the stereotyped images of pristine environments demanded by ecotourism. One important factor in this is the way that ecotourists conceptualize and perform their holiday choices, motivations, and environmental attitudes and behaviours.

Furthermore, the discursive recreation of local environments by global visitors also demonstrates how visitors narrate and redefine their own self-identity. These identities are diverse and contested since ecotourists simultaneously hold multiple ideas of the self. Ecotourism is partly about setting oneself over-comeable challenges that can be used as an indicator of one's sense of initiative, flexibility, and ability to endure hardships. However, for nature and culture in Belize the ways ecotourists talked about their destination choices and environmental attitudes have an impact that stretched beyond merely demonstrating how ecotourists regard themselves and their identities. Rather, such narratives of the self, plus demands for the ideal holiday, are also part of a complex engagement with the global tourism industry.

Ecotourism has started to transform the environment in Belize to meet ecotourist demands for certain kinds of performances. These performances and the business responses to them are interestingly new products of the global tourism industry. The creations of ideal beaches, later presented as natural and authentic, as well as pressures for 'swim-with-the-dolphin' trips and the like, are clear examples. Thus ecotourists can only consume and perform their tourism through a range of options produced by the ecotourism industry. Ecotourism is one specific example of the new, globalized and networked pattern of existence.

List of interviews

 1. David Snieder, Caye Caulker, 12.12.97
 2. Michael and Vasili, Caye Caulker, 12.12.97
 3. Debbie Davis, Caye Caulker, 12.12.97
 4. Shawn Nunnemaker, Caye Caulker, 3.1.98
 5. Christian and Ulrika, Caye Caulker, 20.1.98
 6. Megan, Caye Caulker, 11.1.98
 7. John Colt, Caye Caulker, 18.1.98
 8. Jan, Caye Caulker, 21.1.98
 9. Chandran and Martin, Caye Caulker, 4.1.98
10. Dave, Caye Caulker, 3.1.98
11. Tony, Caye Caulker, 12.12.97
12. Steve, Coral Cay Conservation volunteer, Calabash Caye, 27.1.98
13. Lucy, Coral Cay Conservation volunteer, Calabash Caye, 29.1.98
14. Yolanda Kiszka and Erica Earnhard, San Pedro, 30.11.97

15. George MacKenzie and Katy Barratt, San Pedro, 2.12.97
16. Liam Huxley, San Pedro, 25.12.97
17. Petra Barry, San Pedro, 4.12.97
18. Brie Thumm, San Pedro, 30.11.97
19. Beat Ziegler, San Pedro, 30.11.97
20. Mary Tacey, San Pedro, 23.12.97
21. Pam Stratton, San Pedro, 5.12.97
22. Dan Smathers, San Pedro, 3.12.97
23. Eddie D'Sa, San Pedro, 25.12.97
24. Fabrice Zottigen, San Pedro, 27.12.97
25. Fred, Caye Caulker, 15.12.97
26. Bob Goodman, San Pedro, 5.12.97
27. Barbara Burke, San Pedro, 22.12.97
28. Bob Goodman, San Pedro, 5.12.97
29. Dawn Harbicht, San Pedro, 2.12.97
30. Tim McDonald, San Pedro, 8.12.97
31. Peter Liska, San Pedro, 25.12.97
32. Yvonne Vickers, San Pedro, 30.11.97
33. Jim, San Pedro, 3.2.98
34. Jillian Porter and Patrick Kelly, San Pedro, 31.1.98

Interviews with tour operators and conservation organizations

35. Ricardo Alcala, Ricardo's Adventure Tours, Caye Caulker, 4.1.98
36. Miguel Alamia, Manager of the Hol Chan Marine Reserve, San Pedro, 2.2.98
37. Nicole Auil, Manatee Researcher, Coastal Zone Management Plan, Belize City, 23.11.98
38. Chocolate, tour guide, Chocolate's Tours, Caye Caulker, 19.1.98
39. Oscar Cruz, Belize Diving Services, Caye Caulker, 10.1.98
40. Frenchie, Frenchie's Dive Shop, Caye Caulker, 11.1.98
41. Daniel Nunez, Owner, Tanisha Tours, San Pedro, 30.12.97
42. Alberto Patt, biologist, Hol Chan Marine Reserve, San Pedro, 2.2.98
43. Mito Paz, Director, Green Reef, San Pedro, 2.2.98
44. Melanie Paz, Owner, Amigos Del Mar Dive Shop, San Pedro, 1.2.98
45. Neno Rosado, tour guide, Caye Caulker, 12.12.97
46. Dave Vernon, Toadal Adventures, Placencia, 15.1.99

Notes

1   Rosaleen Duffy conducted a total of 53 semi-structured interviews with ecotourists, and 89 semi-structured interviews with tour operators, academics, environmental NGO representatives, and national government officials in Belize, 1997–2000. The author would like to thank the ESRC for funding this research, grant numbers L320253245 and R000223013.
2   See *www.ecotourism.org/* accessed 25 January 2004. It is interesting to note that many of the businesses listed by the Ecotourism Society as their partners in Belize were visited during the course of this research. Visitors to and users of those businesses were interviewed, as were the owners, tour operators, and so on.
3   See for example *www.quarkexpeditions.com/* accessed 25 January 2004. Quark Expeditions run cruises to Antarctica according to an environmental code, plus a team of expedition leaders, naturalists, and lecturers as part of the package deal.

4   The *Guardian* (UK) 22 April 1998 'Uphill Struggle'; also see chapter 7.
5   See *www.undp.org/sgp/cty/LATIN_AMERICA_CARIBBEAN/BELIZE/pfs5647.htm*, accessed 25 January 2004; *www.sirenian.org/ManateesInBelize.html*, accessed 25 January 2004; and Wildlife Preservation Trust International and the Wildlife Conservation Society 1996: 4–5, for further discussion.
6   Anonymous interviewee. The author also witnessed the activities of sand pirates. Anonymizing interviewees puts the reader at a disadvantage. However, sand piracy is also interlinked with other illicit activities in Belize, especially drug smuggling. Interviewees took a real personal risk when they discussed the topic. For further discussion, see Duffy (2000).

Chapter 5

Surfing: global space or dwelling in the waves?

Rob Shields

> I stand amid the roar
> Of a surf-tormented shore,
> And I hold within my hand
> Grains of the golden sand –
> How few! yet how they creep
> Through my fingers to the deep,
> While I weep – while I weep!
> Edgar Allan Poe, 'A Dream within a Dream'
> (1827, in Poe 1993)

Beaches and waves

This chapter is about beaches and the activities that take place upon them in general – and surfing in particular. As environments and ecologies, beaches are spatialized as margins, as 'free zones', despite being highly organized as sites of work and leisure (Freeman 2002). They are less often thought of as dynamic sites of change and play. Surfing is an activity that exemplifies the dynamic beach as a space of play. Surfing illustrates current debates on the construction of leisure sites as homogeneously 'generic' global spaces of play (Lash 2002).

'Wounded by what love did you die deserted on a barren shore?'

In tourism literature beaches are widely treated as stretches of seaside whose social uses as leisure sites are taken for granted. The image of the 'good beach' is widely connected with sand, for example. And typical beach behaviour is widely construed as some form of leisure. This may be active, such as beach volleyball, running, or collecting shells. Yet a dominant image of beach activity is of sunbathing, prone bodies, of a person reading under an umbrella or shade, or even of a seated group at a picnic or bonfire.

The beach is part of an alternative, 'sunburnt and sand-encrusted history' (Martin 2000: 80) of modernity that contrasts with the official history of governments, wars, and revolutions. The alternative history is one of holidays from our everyday lives, decisions whether to go in swimming or not; anxieties about self-exposure to others and to the sun, and of all the equipment, excuses, and accompanying social games (Shields 1991). On the beach, the body shows itself as a malleable form and flexible performance that aspires to hybridity, underwriting the suspension of social norms and categories of identity. At the same time as identities (such as class) are undermined, differential capacities of bodies for adaptive performances are highlighted (gender, sex, race). It is a form of collective archetype, an 'indeterminate, ambiguous zone between earth and water, raked by sun and blasted by wind...where conflicting elements fight it out for the possession of our souls' (Martin 2000: 80).

Historically, the meaning of beaches is itself mobile. While beaches have always been margins they were not always regarded as spaces of play. Margins are zones of 'unpredictability at the edges of discursive stability, where contradictory discourses overlap, or where discrepant kinds of meaning-making converge' (Lownhaupt-Tsing 1994: 279).

Beaches were historically regarded as dangerous limits, 'surf-tormented shores' at which territorialized or landed social order gave way (as in the early poem by Poe quoted above). Racine's *Phèdre* (1677) presents the classical attitude to the beach: a horrifying space bordering the abyss. The Tahitian travelogue of Louis-Antoine de Bougainville marks the turn to the beach as a hedonistic pleasure zone. *Voyage autour du monde* (1771–72) recounted Tahitian villagers' willingness to barter sex for cargo during Bougainville's brief stop at the island in 1768. This text established the beach as a Dionysian space (cited in Martin 2000: 78; Corbin 1994) and influenced Charles Fourier's conception of a workers' utopia, not to mention an image of beach holidays purveyed by late-twentieth-century resorts.

The continuity and inter-relation of beaches with other coastal morphologies, the dynamic quality of the sand or pebble beach as a site of restless activity, and the erosion or silting of the shore are little mentioned in the social-science literature (although see Bærenholdt *et al.* 2004: chapters 1 and 8). This chapter follows the alternate course and treats the sand beach as a continuation of the fluidity of a body of water. As a shifting, mobile ecology, the beach plays host to various forms of amphibian, animal, and human mobilities. It is a space of coming and going. Not only do tourists arrive as travellers to the seashore, but in-shore fishermen historically departed from beach landings.[1]

Surf's up

Surfing is one activity that typifies certain beaches, or more properly the waves landing on them. Surfing is the art of standing and riding on a board propelled by breaking waves. Long boards are designed to allow one to travel on the face of a breaking wave, while shorter boards with multiple skegs (fin-like keels) are designed to allow one to cut across and even up the face of a wave. Surfing is thus highly dependent on the nature and pattern of incoming waves – their height, undertow, whether they break parallel or at an angle to a beach, and so on. Preston-Whyte notes that the waves themselves are as important as the social and technological aspects of surfing in structuring where surfers congregate and the social forms their interaction takes – with each other, with non-surfing locals, and with the waves.[2]

As a social performance of stamina and fitness, of amphibian and aquatic ability, surfing ties the body to the waves as well as to the beach, to surfboard technology, and to the style trends of surf culture and its representation in magazines such as *Surfing*. As represented in English-language histories and the press, it is bound up with sexuality, in particular a muscular and heroic masculinity, and the competence of the individual on the surfboard. Journalists have recently asserted that the gendered quality of surfing is changing (Supurrier 2002; Kampion and Brown 2003: 153; see also issues of *Wahine* magazine, published up until 2002). However, this is more movie-myth than reality. Surveys such as Preston-Whyte's study in Durban, South Africa, show that fewer than 18 per cent of surfers are female (2002: 314; see also Henderson 2001). What has changed about surfing is that studies show that it now spans generations, with Baby Boomer 'old bullets' returning to the surf on 'guns' – longboards, which require less strength and offer a more forgiving glide compared to the faster short boards developed in the 1970s and 1980s.

While surfing extols the virtues of highly localized beaches where wave formations allow for feats of speed and distance, it is also a global phenomenon, often identified with American culture, but in its origins linked to Polynesia and Pacific-Rim cultures such as Peru, where rock art dating from 3000 BC at Chan Chan is said to depict surfing. Despite being a global media icon of individual leisure and freedom, surfing is hardly universal: one can only surf at an ocean coastline at particular places (where ocean swells have been able to travel for thousands of kilometres without interruption by intervening landmasses), thus restricting who surfs and when. Surfing is a stark contrast with the stasis of sunbathing and other leisure stereotypes. It also presents mostly an individual challenge of a person 'against nature', rather than against another individual or a team (Ishiwata 2002).

From the viewpoint of surfing, beaches are merely a support zone for an even more indeterminate space in the waves, namely 'the tube' or 'pipeline' formed under the crest of a breaking wave, where the highest velocities can be reached as one surfs at an angle down the front of the wave, combining both the momentum of the wave as a moving wedge and the force of gravity. Travelling at a slight angle prolongs the experience, since the crest of ocean waves does not curl or break along the entirety of the wave at the same time. Rather, the crest reaches over the wave to form a horizontal vortex that moves along the length of the wave. The challenge of surfing is to stay 'in the tube', just inside the collapsing vortex. In this region the surface tension of the water is high and the planing performance of a surfboard is optimal. As the wave crest collapses, however, a region of foam is generated that cannot support the board and rider (since bubbles diminish the water's buoyancy). As this almost inevitably catches up with the surfer, 'wipe outs' occur that can occasionally be fatal due to any one of a number of risks (Stranger 1999) – such as the force of undercurrents, being knocked unconscious by the board, or by hitting the bottom.

Conflicts are often noted in the surfing press. 'Locals only' signs mark some beaches. These conflicts also occur between locals – which is to say, between seasoned insiders who resent being crowded, and beginners ('kooks') who surf less decisively but may occupy the best waves. Taking off on a wave demands both timing and location to synchronize the movement of the board with the gathering momentum of a cresting wave as it moves into shore. This pecking order is not only a function of crowding, or rules of turn-taking, but a status order which emerges based on experience and demonstrations of ability and competence which are seen by other surfers congregating at times when the surf is best.

In a sense, the tube is a globalized and mobile place. This is perhaps the most represented setting of surfing magazines. These publications are highly visual, made up of action images, interviews about surfing experiences, pull-out posters, discussions about surfboards, and advertisements featuring yet more action shots. It is as much an emotional place of memory, desire, and fantasies of optimum balance and embodiment. In the tube, a synchrony of forces must occur to maintain the balance, direction and floatation that allows a surfer to keep up with the moving vortex and wave form. Because one is surrounded by rushing water and air, the result is a curious stasis, an illusion of dwelling in a shelter of moving water.

It is the surfing magazines that freeze-frame the fleeting space to capture this moment as a time-out-of-time (cf. Lefebvre on the utopian qualities of this, as elaborated in Shields 1999). With a surfboard and practice, they promise, one can possess this space for a few gravity-defying seconds before being evicted by the waves and ejected back into the human world where we are quickly reminded of how aquatically challenged we are. Other comments can also be read as indicators of a perceived relationship between risk, the sublime, and the speedy stasis of the tube. These qualities make surfing a powerful metaphor of virtual – though not actual – command and control in risky, dynamic environments (of which another example is the web; see Shields 2003). Thus, from its first modern appearance:

> Kahanamoku defined surfing, which had come out of Tahiti a millennium before, as the quintessential post-historical beach exercise. You weren't going anywhere but you rose and you fell in accordance with the rhythms of the ocean; you weren't doing but being. The surfer was at once a compelling biblical parable, riding the flood aboard his personal ark, and a bronzed neo-pagan metaphor in shorts, a being with an unconquerable lusting after the tube, but subject to the ultimate law of the wipe out and death.
>
> (Martin 2000: 80)

'Charlie don't surf?'[3]

Like World War I, surfing as we know it is linked to Edwardian globalization. It first came to the notice of English-speaking publics in 1915:

> Around the time of Gallipoli, while all of Europe was digging itself into trenches, Duke Kahanamoku went to Sydney. He was Hawaiian, a two times Olympic swimming champion, a waterman so strong it took the future Tarzan, Johnny Weismuller, to beat him. On 15 January 1915, Duke left the public baths where he beat his own world record for the 100 yards freestyle, went down to the beach at Manly, carved himself a board straight from a tree and paddled out and caught a handful of waves, finishing with a head-stand. As if by a miracle, a god-like man walking on water, Australia had been converted into a surf-going nation.
>
> (Martin 2000: 80)

Surfing involves riding the swell of waves breaking onto a shallow or gradually shelving shore. The term 'surf', referring to the foam 'soup' that appears at the crests of waves, appears first in the travelogues of seventeenth-century mariners to India. Some indications suggest that the word may originate from Madras (now Chennai). In early references, surf marks the violence of the sea and makes the shoreline a treacherous place for setting out and returning to land.

One of the first European images of surfing is 'View of Karakakooa, in Owyee [Hawaii]', an engraving based on a 1779 watercolour by John Weber, which shows a man paddling out to greet Captain Cook. More learned descriptions appeared by the late nineteenth century (Bolton 1891; Culin 1899). On an early 1870s visit to Hawaii, Mark Twain observed:

> In one place we came upon a large company of naked natives, of both sexes and all ages, amusing themselves with the national pastime of surf-bathing. Each heathen would paddle three or four hundred yards out to sea (taking a short board with him), then face the shore and wait for a particularly prodigious billow to come along; at the right moment he would fling his board upon its foamy crest and himself upon the board, and here he would come whizzing by like a bombshell! It did not seem that a lightning express-train could shoot along at a more hair-lifting speed. I tried surf-bathing once, subsequently, but made a failure of it. I got the board placed right, and at the right moment, too; but missed the connection myself. The board struck the shore in three-quarters of a second, without any cargo, and I struck the bottom about the same time, with a couple of barrels of water in me. None but natives ever master the art of surf-bathing thoroughly.
>
> (Twain 1872/1972)

With the growth of tourism in Hawaii in the 1920s, surfboarding became a common theme in souvenir images and products. *Surfing* magazine, founded in 1960 by John Severson, provided a consolidating mirror to the California beach subcultures. The staff of its art department over the years, such as Rick Griffin, exerted an important influence on the 1960s and 1970s 'supergraphics', record-cover design for bands such as the Grateful Dead and Jefferson Airplane. This work also influenced the visual culture of San Francisco's Haight-Ashbury hippie movement (Kampion and Brown 2003).

It was a Hawaiian, Duke Kahanamoku, who was the first to accept endorsements from a board manufacturer. However, surfing professionals emerged in the late 1950s when Corky Carroll became the first professional to receive endorsements and become a recognizable star, not only on surfing championship circuits but in advertisements. His connection with surfing magazines marks the mediatization of surfing. Southern California photographers such as Dr John 'Doc' Ball, Tom Blake, Robert Johnson, Leroy Grannis, and Don James played an important role in the representation of inshore waves, surfing iconography, and the standardization of surfboard 'routines' that could be compared, and thus scored, in competitions (Booth 1995).

Surfboard design also evolved from the 1930s to 1960s from carved and shaped plank boards to shorter polyurethane foam surfboards that allowed non-fading, bright pigments and silkscreen prints that artists and designers used to create eye-catching graphic designs.[4] As a mass-produced, youth-oriented sport associated with risk and sunburnt, toned masculine bodies, the mass-produced surfboard was an ideal vehicle for florescent graphics that quickly influenced advertising and product design. Surfing was popularized in escapist films such as *Gidget* (1959, directed by Paul Wendkos; discussed in Rutsky 1999) and in the musical lyrics of internationally marketed bands such as the Beach Boys (Blair 1985).

This band was only one of hundreds of garage bands that emerged in California between 1960 and 1963 and who created a heavy-reverb staccato sound on the low strings of Fender electric guitars and Fender Showman amplifiers. Dick Dale was primarily responsible for developing this so-called 'surf sound'. The music was non-threatening, white, and dance oriented, even though surfing was widely understood as a form of resistance to regulation, an escape from Cold War (see Booth 1994). Jan & Dean wrote and recorded hit titles such as 'Surf City', 'Little Honda', and 'Sidewalk Surfin'. Unlike most of Dale's imitators, the Beach Boys introduced vocals in harmony and sought to exemplify a Southern Californian teenage lifestyle in the lyrics that drew on local youth vernacular, including the importance of the car as the key form of transport to the beach:

> I got up this morning, turned on the radio,
> I was checkin' out the surfin' scene to see if I would go
> And when the deejay tells me that the surfin' is fine,
> That's when I know my baby an' I will have a good time.
> I'm goin' surfin'...
>
> (Beach Boys 1961)

'Surf' music was unique in that it introduced the United States to a cultural form based on a youth subculture and sport whose geographical restriction meant that only a small proportion of the population could actively participate. It also marked the emergence of Los Angeles as an important production centre in the music industry. It had enduring impact – even the satirical cartoon *The Simpsons*, Bart Simpson echoes its vernacular with his favourite but unthreatening surfer expletive 'Cowabunga!' By 1965, however, most bands had shifted to hot rod and car songs. Motown, together with the Beatles and the 'British Invasion', challenged musical taste.

Attempts have also been made to use surfing as a tool of diplomacy, by pointing out its cross-cultural popularity as a shared element of Pacific Rim culture. However, its equivocal status is captured in a 1992 *New York Times* article that discussed surfing as a cultural tie between the United States and East Asia but ironically used an illustration from *Apocalypse Now* (1979).[5] Indeed, the technology of surfboards has also followed defence-industry developments in materials and theories of hydrodynamics. In *Apocalypse Now*, troops surf while a village is strafed in the background and battle is engaged to capture a part of the shoreline. Building on the legacy of the Beach Boys' 1963 hit single 'Surfin' USA', surfing is constructed in the film as quintessentially white, American, and as an aggressive appropriation of a resistant activity. Rather than being a 'shared' cultural activity, then, the film presents surfing as a marker of leisure privilege and masculinity, suggesting cultural colonization and the take-over of an indigenous activity by white commercial interests.

The linkage between surfer masculinity and the beach as a locus of youth subcultural resistance to adult prerogatives is mirrored in the aggressive side of surfing iconography as an extension of military invasion and occupation. A recent James Bond film (*Die Another Day*, 2002) tips its hat to *Apocalypse Now* by opening with its protagonist surfing on heavy waves – surfer-defying 'crunchers' folding over themselves – into a bunkered and booby-trapped shore that the audience is to imagine is North Korea, yet has more resemblance to World War II images of Omaha Beach or Dieppe.

Kinaesthetics

The kinaesthetics of surfing combine both the stability of balance and relative speed keeping up with a wave front, and mobility, in that the surfer – in fact the entire assemblage of human, board, and wave – are simultaneously in headlong flight and minutely and dynamically adjusting to maintain the form of the wave and the surfer. This mobility is also stasis-like in the sense that it is a vector without a goal – other than perhaps its own dynamic equilibrium. The mobility of the surfer does not make one think of a destination – whether the beach or an itinerary beyond. In the semiotic sense, surfing has no 'meaning' beyond itself alone. Even as mediated in magazines, surfing images are presented as full-page spreads, printed to the borders of pages without text, inviting the reader to transcend the narrative line of any given story and fluidly imagine themselves as a virtual presence in the orderly disorder of the tube.

Surfing is play in Huizinga's sense of a non-utilitarian and unproductive activity outside the space of everyday social interaction, and regulated by arbitrary and contingent conventions and subcultural knowledges (1950). It takes place in a space set aside for it. 'Doubt must remain until the end, and hinges upon the denouement...An outcome known in advance, with no possibility of error or surprise, clearly leading to an inescapable result, is incompatible with the nature of play' (Callois 2001: 7).

Is surfing an example of changes in play? Lash argues that leisure is altered by disembedding it into an 'anyplace-whatsoever' (a term he quotes from Deleuze 1992) or what Lash calls a 'generic space' (after Koolhaas and Mau 1997). The essence of play is rhythmic movement but one that seems to have no effectiveness:

> Play in the information society takes place in a generic space. Generic spaces are disembedded spaces that could be anyplace...They are 'lifted out', so to speak, from any particular context and could take place in any context....
>
> (Lash 2002: 161)

Is the tube a generic space? Is it a global space of play? Its mediatization in film and in surfing magazines might suggest so. Surfboards are a technology of mobility that extends the automobile as an icon of youth identity, and derives from advances in military aviation. Surfing is repetitive, rhythmic, and not definitively located in a single spot on land – a quality that fits well with its image as a masculine lifestyle that emphasizes dedication and a global pursuit of the very best places to surf (Taylor 2002).

However, surfing defies the construction 'generic' precisely because of qualities that are both performative and unpredictable. Its motions are too artfully composed to be merely a matter of chance, the directions the surfer takes on a large wave seem random, and change suddenly. But the grace of the surfer seems to defy the force of nature.[6] Surfing is play, but the performativity of surfing is more closely connected with the body adapting to local conditions to achieve a mimetic action that allows the surfer to stay in control of the board. Surfing magazines emphasize images rather than text, and routines are described in specialized terms in order to explain what might be involved in achieving the pictured performance.

The embodied experience of the body-board-wave assemblage in action produces transcendent emotions or 'flow' experiences (Csikszentmihalyi 1990). Surfing mobilizes feelings. But there is a space for personal memory even in looking at the magazines as one reacts to the surfers' body positions and expressive postures that are possible only with speed and centrifugal force. There is an impulse to mimic the surfers, which one may only detect when one notices the tension in one's face or body after the fact. These postures are generic, but personal memories are inflected in myriad ways related to individual biographies.

Surfing is an obsession because each wave is a new challenge, a different thrill. And each beach offers a different surf, some being famous for their 'crunchers', others for their 'bomboras' (Australia) and 'heavies' (Hawaii). Similarly, there are famous beaches such as Makaha in Hawaii, Huntingdon Beach in Orange County, California, or Bondi Beach near Sydney, Australia. What they offer is a 'moment', in Henri Lefebvre's sense, a temporal utopia embedded in – not 'lifted out' of – the mundane (see Shields 1999).

Notes

1   See Acheson 1988; Ellis 1986; Davis 1991; Matthews 1993.
2   Preston-Whyte mapped surfers' knowledge of wave types and location of rip currents on beaches near Durban, South Africa (Preston-Whyte 2002: 317). The importance of this environmental awareness, and the perception of the best wave-forms as a limited resource, has also been noted as both the basis of conflict between surfers (Weseman 1998, cited in Preston-Whyte 2002) and environmental activism (Ward 1996; Beder 1991).
3   Attributed to John Milius, screenwriter, Francis Ford Coppola, and Michael Herr. The character Colonel Kilgore (played by Robert Duvall) in Coppola's film *Apocalypse Now* makes this remark when asked why he is confident his American air cavalry unit will oust Vietnamese Communist troops from a salient on the beach (1979; see Hansen 1980; Lyons 1997)
4   By graphic artists such as Hap Jacobs, Mike Salisbury, and designer Dave Rochlen.
5   An exhibit entitled 'Surf Vietnam' by the artist Simon Leung (curated by Tyler Stallings in Summer 1998) at the Laguna Art Museum in Laguna Beach, California, brought together Vietnamese immigrants, veterans, and surfers to explore the legacy of American–Southeast Asian violence.
6   My comments here are inspired by Jean-Paul Sartre's 1948 discussion of Alexander Calder's 'stabiles' or sculptures with moving, wind-blown elements and armatures. I am indebted to Anne Galloway for this source.

Part 2

Performances of global heritage

Little England's global conference centre: Harrogate

Viv Cuthill

If Harrogate ever had life, it was dead now.

(adapted quote from *Withnail and I* (in Murphy 2000))

Introduction

Harrogate has always been a place for embodied play, a playground for those with a certain wealth. From the discovery of its first mineral spa waters in 1571, people have visited Harrogate to immerse themselves in a range of activities. Its first incarnation was as a place promoting health for the sick or the wealthy to visit; a place to recuperate and play. Harrogate really developed following the arrival of the railway in the town in 1848, enjoying its heyday as a spa during the Victorian era up until World War II. However, after the war the demand for spa treatments faltered. With a large accommodation stock and a population relying on service industries, Harrogate began to develop as a conference town. For those living in the town or near it, the facilities provided for visitors meant that Harrogate was also a local place to play for residents and day visitors.

In this chapter I explore how Harrogate, has been socially and culturally constructed through global flows of people, objects, finance, technology, and signs. I consider how Harrogate has been produced and reproduced through these flows, and how the speeding up and disjuncture of flows are leading to changes in the lived experience of Harrogate, resulting in a local crisis in place-identity.

Harrogate, spa town, international conference centre, and 'Gateway to the Yorkshire Dales', is located in North Yorkshire, England. Geographically it is located centrally along Britain's north–south axis, equidistant from Edinburgh and London. Visitor statistics for the Harrogate District show that in 2001, £142.1 million was earned from business tourism and £51.2 million from holiday visitors, with an estimated 335,000 business visitors attending conferences, exhibitions, and trade fairs in Harrogate (Harrogate Borough Council website: July 2003). In terms of conference tourism earnings, Harrogate is ranked third in England behind London and Birmingham (Neesam 2001: 104).

The 2003 Harrogate District Guide focuses on the appeal of Harrogate as a conference and exhibition venue, as a floral destination, and as a place for exclusive shopping and alfresco café society. Harrogate's main attractions are the wide range of events staged there every year, whether business, floral, antiques, arts, or agricultural. It is a place to which the epithets 'floral town' and 'antiques centre of the north' are often applied, its other main draw being its spa-town heritage and its range of amenities, in particular places for cosmopolitan eating and drinking.

Part of Harrogate's attraction for the visitor is the 'being there'. Many people visit Harrogate today with the perception that it is an exclusive, elegant place with a spa heritage and service traditions continuing into the twenty-first century. However, a 'new' Harrogate, a new place to play, is emerging, of a café-bar culture that has led to claims that it is becoming a 'Jekyll and Hyde' location: elegant Yorkshire town by day, drinking and club town – creating a 'yob culture' – by night. The question of temporality in a 'place to play' is interesting to explore. Place-myths of Harrogate are now contested, as different performances of place occur by night and by day, resulting in disjunctures between the preservation of Harrogate's 'Little England' heritage and new forms of play developing in the town.

Global place

A key question about place is whether, as a result of the impact of processes of globalization, places are becoming 'placeless' – that is, losing their individual distinctiveness (Relph 1976). Does the increasing internationalization of economic structures, the growing power of multinationals and corresponding weakening of other institutions, the flow of international migrants and tourists, instantaneous global communications, and time-space compression lead to greater homogeneity of place, or do places still have distinctive identities and do they generate unique senses of place?

Tourism development creates specific images (Kotler *et al.* 1993) and generates certain global flows in and out of places, but these are not the only flows and images that construct places socially. Squire (1998) observes that tourist destinations are social and cultural constructions whose meanings and values are negotiated and redefined by diverse groups, and mediated by factors often related only tangentially to a particular tourist setting. Places do not just change as a result of the development of tourism *per se*, but respond to social, cultural, economic, and political change in a globalizing world. Therefore I explore a re-conceptualization of place and space within a distinct tourist destination.

Bauman calls for the use of metaphors of 'fluidity' and 'liquidity' to help us grasp the nature of the present phase in the history of modernity (Bauman 2000). To facilitate an understanding of the impacts of diverse fluid mobilities on place, I will examine Urry's proposition that:

> Places can be loosely understood therefore as multiplex, as a set of spaces where ranges of relational networks and flows coalesce, interconnect and fragment. Any such place can be viewed as the particular nexus between, on the one hand, propinquity characterized by intensely thick co-present interaction, and on the other hand, fast-flowing webs and networks stretched corporeally, virtually and imaginatively across distances. These propinquities and extensive networks come together to enable performances in, and of, particular places.
>
> (2000:140)

In order to explore this further I adopt the framework proposed by Appadurai (2000: 230–31), who suggests that the new global cultural economy needs to be understood as a complex, overlapping, disjunctive order, its complexity arising from certain fundamental disjunctures between economy, culture, and politics. His framework for exploring these disjunctures examines the relationship between five dimensions of global cultural flow: ethnoscapes, mediascapes, technoscapes, financescapes (financial interactions), and ideascapes. These scapes relate to different sorts of actors – from the global to the individual – and Appadurai proposes that these landscapes are the building blocks for 'imagined worlds' – multiple worlds constituted by the historically informed imaginations of people and groups around the globe. When we examine the very local case of Harrogate through this framework of disjunctures, we can see how different 'imagined Harrogates' exist in ways that either reinforce, contest, or subvert traditional images of place.

While Castells (2000) proposes that in the network society places become disembodied from their cultural, historical, and geographical meaning (the 'space of places') and meaning is instead found in global flows of images (the 'space of flows'), Massey (1993, 1994) has argued for the need to adopt a 'global sense of place'. This is a progressive sense of place that is not a bounded, coherent entity, but instead a meeting-place constructed out of a particular constellation of social relations, meeting and weaving together at a particular locus. Massey proposes that places can be imagined as articulated moments in networks of social relations and understandings, where a large proportion of those relations, experiences, and understandings are constructed on a far larger scale than what we happen to define as the place itself. She maintains that this allows a sense of place that is extroverted, includes a consciousness of its links with the outside world, and integrates in a positive way the global and the local.

Hetherington (1997) identifies the importance of the materiality of place in terms of mobility. He proposes that places are not just attached to space, but are diasporic: they travel with us and are articulated through mobile objects. Thus we can consider how mobile objects that represent Harrogate as a place travel across multiplex networks and thereby create a global sense of place. However, Crouch also points out that the character of space in tourism practice is a combination of the material and the metaphorical. The body encounters space in its materiality, but space is also apprehended imaginatively, 'through our own engagement, [and] imaginative enactment' (2002: 208). Through my analysis of embodied performances in and of Harrogate's space, I show how a shifting sense of place is created through material and imaginative enactments of this tourist destination.

Imagining Harrogate

Harrogate is an interesting case study because it has always been a focus of tourist activity; its life as a tourist destination has made the place. The town has been shaped by the practices of spa tourists and latterly by the practices of conference tourists, and the qualities and ambience of the place live on through its spa associations. Senses of place are enacted through performances of heritage, and Harrogate's spa heritage is central to place-myths of the town. This section will examine the various ways in which place-myths of Harrogate have been created, and how corporeal, virtual, and imaginative experiences of Harrogate contribute to senses of place. It is important to identify how these senses of place have been produced, contested, and reproduced and changed over time.

Until recently Harrogate had a fairly stable set of place-images relating to its identity and the way it was imagined by different communities. A well-documented contribution to the place-myth of Harrogate from the days of its popularity as a spa is when Agatha Christie chose it as a place to hide away when she went missing in 1926.

The author of a recent book outlining this story explains Christie's success in not being immediately discovered thus:

> Within the rigid British class structure of the time, Harrogate's prosperity as a spa town depended on its assuring comfort, excellent service and total discretion to its wealthy and influential visitors. Harrogate's status as the pre-eminent hydropathic centre of Europe arose from its eighty-seven mineral waters [springs] and the first-class service of its hotels and shops.
>
> (Cade 1998: 111)

Immediately we can see the features of Harrogate that made it appealing at the time: comfort, discretion, and first-class service. These intangible features of place were portrayed in the film *Agatha* (1979), set in Harrogate and starring Dustin Hoffman and Vanessa Redgrave. They continue to be expected by visitors today based on the reputation of Harrogate and the myths that circulate about the place, now reproduced in Cade's work on Christie's stay in the town, thus demonstrating how myths reappear and perpetuate themselves through a range of media.

This middle-class, affluent, respectable town in Yorkshire is seen as a desirable place to live as well as to visit, and some of the particular features that made the town attractive in the 1970s are summarized by Roy Hattersley:

> Harrogate and its people go well together. The town possesses all the attributes of the determined middle-aged ladies who frequent its teashops and patronize its not-quite-chic boutiques. It is moderately well preserved. It is deeply conscious of the need to keep up appearances and remain, at least in public, neat and tidy. It is insulated from the uglier aspects of life in the West Riding. And, above all, it is self-satisfied. Carefully manicured gardens, tulips blooming clean and upright, grass on Stray [a substantial grassed area near the town centre] crisp and perfect. Its motto should be Do not Disturb.
>
> (Hattersley 1976: 91)

In this description Hattersley's perception of people and place is deeply intertwined. At the time an image of Harrogate was socially constructed through the performances of middle-aged ladies, and even nature joined in and behaved appropriately. Its virtues are discipline and good order, neatness and sobriety.

Perhaps because of the strong place-image, there are various myths that circulate about its location. It is a place in the North of England and yet it does not seem to be 'of the North' (see Shields 1991, on the 'north' of England). As Bryson comments:

> You would never guess that a place as prosperous and decorous as Harrogate could inhabit the same zone of the country as Bradford or Bolton, but of course that is the other thing about the North – it has these pockets of immense prosperity, like Harrogate and Ilkley, that are even more decorous and flushed with wealth than their counterparts in the South.
>
> (Bryson 1995: 219)

Harrogate Borough Council maintains a series of venues that were formerly frequented by spa visitors for treatments and entertainment. These are grand iconic Victorian buildings and floral open spaces that together present the image of a spa town and a high-quality environment alluded to by Cade, with names such as the Royal Baths (which house the famous Harrogate Turkish Baths) and the Royal Hall. Spa visitors would promenade in areas such as the Valley Gardens with its flowers, walks, sun pavilion, and bandstands. High-quality hotels carry famous names and associations both locally and nationally, such as the Majestic and the Crown. All of these public and private buildings and spaces were created for tourist performances, and together they constructed the town centre and the distinct image of the town. Visual images of Harrogate's iconic buildings and floral spaces circulate and recirculate through a variety of media: postcards, photographs, paintings, TV series, websites, tourist brochures, and so on. Seeing and visiting these iconic buildings and spaces can be part of the performance of the Harrogate experience.

Images of Harrogate have been circulated, recycled and recirculated, carried along mediascapes in expected and unexpected ways, to produce and reproduce complex narratives of place. These are generally linked to the ethnoscapes of tourists and other people moving through the place or experiencing it imaginatively through pictures and the narratives of authors. The flows of images at this stage complement official council ideascapes of place. They also serve to create expectations of the appropriate performance of place.

The imagination of a place is not just about images and descriptions, however, but also about enactments of being there, forms of eating and drinking there, and the branding of objects that come to represent the place. In the internationally acclaimed books about his life as a vet in the Yorkshire Dales, James Herriot portrayed Harrogate as the fictitious town of Brawton. In reality Herriot and his wife used Harrogate as their place of escape on their days off. For them it was a 'weekly haven' and 'exhilarating leisure-place'. He recalls the gentle air of Victorianism and the town's clean beauty. In a clear demonstration of the importance of performance in the experience of place he recounts his Harrogate routine of thirty years: 'Lunch at Betty's or Standings, shopping in the afternoon, tea at Betty's, then the cinema…and finally a late meal at Louis' restaurant' (1979: 158–62).

'Being there' and sensuous experiences are central to the distinctive practices of performing Harrogate. There are various ways in which place-myths of Harrogate lead the visitor to expect a sensuous experience: by promenading through the town smelling the fresh Yorkshire air and the flowers; by steaming in the Turkish Baths; by 'taking the waters' – sampling, perhaps ironically, the repulsive taste and smell of the sulphur water; by feeling, tasting, smelling, and purchasing up-market goods in the shops; and by participating in a high-quality service experience in hotels, cafés, and restaurants.

Some iconic local products that should be smelled, tasted, and consumed are Taylor's of Harrogate teas and coffees, Farrah's Harrogate Toffee, and the confectionery and baking produced by Betty's Café Tea Rooms. Just as iconic buildings and spaces carry visual images and associations beyond the boundaries of the town, mobile iconic food and drink objects produced in Harrogate also carry images of quality and desirability. These objects have a global reach and play a key role in the social construction of place. As Lury notes: 'the career or biography of objects – their emergence, movement, or salience in relation to other social practices, including but by no means confined to those of economic exchange – also has the capacity to influence not only the preferred destinations of the tourist but also the nature of the tourist practices undertaken once the tourist arrives' (1997: 77).

In the time that Herriot was visiting Harrogate you needed to be there to have the sensuous experience of place, but nowadays these products also create a global web of consumers of Harrogate experiences of place who can sensuously experience Harrogate without being present. Taylor's of Harrogate has been producing high-quality teas and coffees since 1886, their products sourced from the former British Empire. Although they seem like exclusive, locally-produced goods, the teas and coffees are to be found in supermarkets, on a range of websites,[1] and are used in cafés, bars, workplaces, and hotels across the globe as markers of 'Little England' quality and distinction. Every piece of packaging, with its prestigious-looking imagery and lettering, carries the name Taylor's of Harrogate. As Lury proposes: 'environments or contexts-of-use are designed into the object-ness of the objects of global cosmopolitanism through techniques in which culture as place is put to work, standardized, operationalized, and redistributed as a construct or property of (some) objects' (1997: 85).

Thus these place branded objects contribute to the global circulation of commodified representations of Harrogate and reflect and reinforce a global sense of place. The reproduction of Harrogate tea and coffee experiences away from the place itself may seem a dilution of the uniqueness of Harrogate. However, Molotch argues that while we can see the same goods and services in so many places, it does not mean that these 'anywheres' can innovate, initiate, and distribute these omnipresent things (2002). Harrogate is the unique producer of these cultural goods.

While Harrogate teas and coffees have been sold globally for many years, currently it is the greater reach of these products flowing via global technoscapes and financescapes that enables these mobile hybrids to travel and be experienced independently of place. It is the speed with which they travel, and the greater web of individual and corporate aesthetically reflexive consumers who are able and willing to access these products at the touch of a button, that multiply and further reproduce virtual and imaginative experiences and images of place.

With increasing personal mobility the influence of ethnoscapes should not be under-estimated either. Migrants into and out of Harrogate are one source of consumers, and as Molotch notes, tourists often buy souvenirs unwittingly when they get home, selecting goods that contain images and ideas they can associate with their visit to recapture some of the exotic milieu and feeling of place (2002: 677).

Betty's is an eating and drinking institution located in the town centre that is a requisite part of the Harrogate experience for visitors and locals alike. Betty's provides hand-made, locally produced foodstuffs and Taylor's tea and coffee. Today it is a symbol of the continuity of prestige practices in Harrogate. A Swiss confectioner, Frederick Belmont, who took a wrong train and ended up in Harrogate, founded Betty's Café Tea Rooms in 1919. The town reminded him of home, and he decided to stay. It is interesting to note that this quintessential English tea-shop is a cultural hybrid.

Belmont is one of a long line of international entrepreneurs, chefs, and waiting staff who have chosen to settle in Harrogate. Thus the in-migration of service-sector workers, bringing their particular cultural attributes and skills, has created an extensive sector of cosmopolitan restaurants and hotels that is somewhat unexpected in a small Yorkshire town.[2] Visiting the restaurants is central to the sensuous place performances of many residents and visitors. Harrogate has always been open to interconnections with the outside world and subject to global flows in and out of the place. However, to follow Harvey (1989), it is the compression of time and space that is leading to current change in Harrogate. While Harrogate has been influenced by global food and service styles for many years, the variety and choice of these outlets has increased since the 1980s, supported and fuelled by the growing conference trade in the town. The influence of global flows travelling along ethnoscapes and financescapes have come together to produce (and simultaneously demand) a particular eating, drinking, and service culture that is characteristically Harrogate, but not 'of' Harrogate, adding to the established place-myths and modifying perceptions of place and the nature of performance.

Events bring groups of people together to participate temporarily in an embodied place experience. Massey (1993) suggests that we should think of each place as a meeting place with a mix of its own particular activity spaces, social relations, and social groups. This intersection of meetings and the resulting encounters lead to new effects and processes that contribute to the uniqueness of each place (see chapter 16 on the effects of meetings and intersections in Dubai). Harrogate has some long-standing annual events that have helped to reinforce and cement its place image, such as the flower shows, antiques fairs, and the Great Yorkshire Show. Business events also contribute to place-myths, with the range of conferences and exhibitions that take place throughout the year leading to numerous criss-crossing and intersecting social flows. Links are made between the identity of the place and the identity of the groups that inhabit Harrogate on a temporary basis, each event creating different senses of place for different groups and adding to Harrogate's global sense of place.

These temporary visiting business populations have effected certain physical changes in the sense of place, the place experience, and the place-myths of Harrogate. In 1988 Alan Bennett filmed *Dinner at Noon*, a documentary for the BBC about the Crown Hotel in Harrogate; he reflects on the experience in his autobiographical work *Writing Home*. He conjures up some traditional images of Harrogate: 'town of teashops, a nice run-out from Leeds'. But it is Bennett who sounds a change in the atmosphere in one of Harrogate's exclusive hotels. He notes that in the late 1980s Harrogate had become a place for 'a "Leisure Break", a conference, a mecca for the businessman' (Bennett 1994: 42). He observes that business meetings are a new innovation in the hotels: 'Reluctant though I am to admit it, I can see that with their conferences and camaraderie and their leisurewear it's business people like this who are banishing class from hotels and elsewhere...the world has changed and maybe it's the businessmen who've changed it' (Bennett 1994: 59–60).

New place-myths

I have examined traditional place-myths of Harrogate and have argued that these old myths become infused with new ones as change impacts upon place. In the case of Harrogate I propose that economic change has brought about cultural change, and that this cultural change is reflected physically, imaginatively, and performatively in place. I will now further develop the place-myths of Harrogate by examining contemporary Harrogate and seek to show how an environment of economic and cultural change, and an increase in global flows, lead to major disjunctures in senses of place.

Local authors writing about Harrogate tend to illustrate its development and appeal through photographs as much as by the written word (see Neesam 1983, 1999; Jenkinson 1998). Its traditional appeal is linked to its architecture, its use of buildings, and the continuity that the physical environment represents. I will relate current wider changes in the town and its image to certain key physical changes in architecture and the use of buildings that act as symbols of change.

The opening of the £34 million conference centre in 1982, the Harrogate International Centre, is one key change. This newcomer to the town's iconic buildings gives the appearance of a spherical spaceship having landed in a northern English town (see figure 6.1). The arrival of a McDonalds restaurant in the town centre in 1985 is another key change. Its development was strongly contested and objected to locally, and the letters page of the *Harrogate Advertiser* still features complaints about the multinational's perceived detrimental effect on the town. The loss of the town market hall, pulled down and replaced by the Victoria Gardens Shopping Centre in 1992, is another key moment of physical change. Victoria Gardens' visual impact is summarized by Bill Bryson: 'the new shopping centre is just heartbreakingly awful, the worst kind of pastiche architecture – a sort of Bath Crescent meets Crystal Palace with a roof by B&Q...a kind of Disneyland meets Jolly Olde England' (1995: 217). The redevelopment of the Royal Baths in 2001–02 as a drinking venue is another key change discussed below.

What these buildings represent is the inflow of international capital, the march of national and international multiples, the rapid increase in the provision of standardized consumption experiences, and Harrogate's entry into McWorld. They represent direct economically driven challenges to key markers of status and identity in Harrogate. Independent exclusive shops, restaurants, and hotels have provided the first-class service expected since the spa town's heyday, whereas fast-food outlets and chain hotels, restaurants, café-bars, and stores have begun to erode the service culture, exclusivity, and uniqueness of 'Harrogate'.

Prior to the arrival of the conference centre, the conference business it now services was absorbed into the existing spa buildings, hotels, and anonymous exhibition halls. The Harrogate International Centre now dominates the lower end of town and the conference trade drives local government decisions about the town's further commercial development. Neesam contends that the controversial development of the conference centre – with the town council as developer – has 'poisoned' public life in Harrogate, by draining away public funds that might otherwise have been used for the restoration of historic buildings and spaces (2001: 105).

The globalization of the world economy has led to a corresponding increase in business travel, with the transnational nature of international business creating a need for those working cooperatively at a distance to also experience corporeal co-presence. Urry draws attention to the importance of conference travel to business networks, as the social and business interactions that he defines as conference 'meetingness' generate stronger ties for groups of people through the different types of physical sociability they experience together (2003b). Williams and Shaw highlight the fact that visitors to conferences in Europe – fuelled by company expense accounts – spend 2–2.5 times

Figure 6.1   Harrogate International Centre

Figure 6.2   Royal Baths, Harrogate

more per day than the average tourist with meals out, entertainment, and shopping accounting for nearly 45 per cent of delegate expenditure (1994: 222). For many, the conference location becomes a liminal zone where workplace rules cease to apply; therefore they expect to have an especially good 'unregulated' time in Harrogate.

Current dominant changes in the social and cultural construction of Harrogate have resulted from the rapid development of new drinking places since the late 1990s, designed to appeal to the conference trade as well as local users (see endnote 2 below). Traditional pubs have been turned into theme bars with names such as the Rat and Parrot, and major brewers have moved into the town with pubs as well as café-bars and theme bars (see Chatterton and Hollands 2003, on changes in urban nightlife in the UK). This new drinking culture is creating place-myths that are challenging and contesting the stability of the imaginary Harrogate that extols the virtues of a comfortable, middle-class, northern English town with its notions of respectability and sobriety, particularly at night.

Café-bars and restaurants now abound in the streets adjoining the Harrogate International Centre, every one keen to acknowledge its proximity to the Centre with pedestrian-management practices that rechannel visitors along these streets. The 'Rat and Parrotization' of Harrogate continues apace, with the council eagerly encouraging a host of new ambient bars and cafés – places usually combining at least two of the functions of bar, restaurant, and club with DJs and chill-out zone. New hotel development is also being encouraged with the 'stylish' Hotel du Vin boutique chain opening in September 2003. One seal of approval for Harrogate as a profitable international location came with a Spearmint Rhino Gentleman's Club opening in 2001. As ever, contributors to the letters page of the local paper were hugely hostile, this time to a lap-dancing venue in the town centre.

With its compact town centre Harrogate has become a place for partying in Yorkshire. Coach groups from northern cities travel for a night out, stag and hen parties choose it as their preferred location, and large groups returning from football and racing divert to Harrogate. It has reinvented itself and become a desirable place to visit for new groups of consumers, who participate in alternative embodied experiences of place. Malbon (1998) demonstrates how the practices of clubbing are rooted in the social life of the city. Harrogate has adopted city practices that appeal to urban consumers, that encourage new flows along ethnoscapes, and create new global senses of place, an urban nightscape.

The unintended consequence is that Harrogate now has a split personality, the 'Jekyll and Hyde' syndrome reported by the *Harrogate Advertiser*. Local hotels struggle to reconcile the contradictions on their websites: 'Harrogate is sophisticated and grand, yet it still retains its small town homely charms. It can be serene and picturesque, but is also lively enough to be one of the premier conference venues in the world and has a vibrant nightlife' (Kingsway Hotel: August 2003). Local opinion is divided, with many sharing the views of a contributor to a national radio programme: 'Harrogate by day is beautiful, by night it is a den of drunken debauchery, by dawn it is awash with urine, vomit and chewing gum' (e-mail contribution by listener to debate about impressions of UK towns and cities on BBC Radio 5 Live, 16 August 2003). These new, unofficial place-myths of Harrogate are leaking out across the wired world. They hit the headlines in October 2002 when the BBC TV programme *Inside Out* exposed some of the problems created by the town's night-time culture. Websites promoting nightlife options provide reviews of the venues in Harrogate (see *www.northnights.co.uk*, for instance). However, the official council website (*www.harrogate.gov.uk*) and tourist promotional materials continue to promote the image of 'traditional' Harrogate.

The iconic Royal Baths building (see figure 6.2) is a fitting symbol for the current contested place-identity of Harrogate. It was of central importance to the Victorian spa in providing health treatments and entertainment. But it had, with the exception of its Turkish Baths and the tourist information centre, lain empty and virtually disused for several years. Located just to the side of the conference centre it has become a building of incongruous mixed uses. The Royal Baths retains its austere frontage, with the tourist information centre at the front and the council-maintained Turkish Baths on one side. Housed within this impressive building are now a range of new national chain drinking venues: Bar Med, Wetherspoons, and Revolution Vodka Bar, two of which become nightclubs after 11pm. The three bars have a capacity of 3,000 and represent the image of the 'new' Harrogate sitting uncomfortably alongside – and in a way *in*side – 'traditional' Harrogate (see chapter 3 on the nightscapes of Cyprus, and chapter 12 on those of Barcelona).

New myths of place have not displaced the old. Harrogate provides different place experiences at different times of day: 'traditional' Harrogate by day, 'new' Harrogate by night. In 2000 a journalist, Bill Murphy, recorded his reflections on Harrogate when he travelled around what he described as the 'decaying heart of England'. Arriving in the afternoon he was unable to find a pub near the railway station and noted the ageing population in the town. He found a theme bar and the service he received there led him to conclude that 'essentially, while Harrogate is great to look at, the once-traditional values of the people of the county of Yorkshire have vanished to be replaced by zombies whose personalities are as vacant as those you might find anywhere else' (Murphy 2000:188). He states that if Harrogate ever had life, it was dead now. For Murphy this was a moment of intensely thick co-present interaction and judgement, where his sense of place was created by imaginatively intertwining a place with its people, in particular the service experience in a theme bar, which creates his sense of placelessness. In a fast-moving world, Murphy feels able to summarize a place after a fleeting visit. This is how the world is made up for many mobile people: brief snapshots of place-experiences as they move in and out, and look quickly for referents and signifiers to confirm their assumptions. Chaney's adoption of the 'glance' as metaphor rather than the 'gaze' is a useful way of considering how global citizens reflexively engage with mobile places and how, in Chaney's words, their 'haphazard consumption can be described as a form of play in public spaces' (2002: 200). Ironically, at the time that Murphy was considering Harrogate to be a town of the living dead, it was more of a play-place by night than it had ever been. Murphy simply chose the wrong time of day to arrive for his Harrogate experience – although for other age groups it would be precisely the right time of day.

There are further ironies. While spa Harrogate and many of its traditional values are in decline, the prestige goods of Harrogate, the mobile food and drink objects, are in the ascendant, disconnected physically from place yet carrying messages of traditional Harrogate values and its spa town status around the globe. In particular the revival of Harrogate Spa Water in 2002 is drawing world attention to Harrogate. In that year Harrogate Spa Water won a gold medal at the Berkeley Springs International Water Tasting Festival in America. In August 2003 the company secured a contract to supply water to major airline catering companies, thus ensuring that in the non-place of the aircraft (Augé 1995), global passengers will have a distinctive local experience of place. In an article for *Observer Food Monthly* (August 2003) William Leith visits a sommelier to taste the highest-quality waters in the world; Harrogate Spa Water is the English representative. The image projected by the water is a modern one: the bottle is glass, in a classic style, the label is black and silver, minimal and modern, and depicts figures in gymnastic poses, thus reflecting a 'new' Harrogate.

Although Harrogate's spa heritage would appear to have been sidelined in favour of the global conference market, the spa itself may not be in terminal decline. There is currently a spa revival across Europe, reflected in the £22 million redevelopment of Bath Spa. Thermae, the Dutch developer behind the Bath Spa project, is reported to have the 'long-neglected spa town' of Harrogate in its sights (*Guardian*, 10 May 2003).

Conclusion

To return to Urry's (2000) notion of place as multiplex, global flows of people, objects, finance, signs, and technology have always moved in and out of Harrogate. As a long-established tourist destination its traditional senses of place have been constructed corporeally, virtually, and imaginatively, in and of these global flows through consumption practices, embodied performances of place, and the consumption of the visual attributes of place and its people. In a fast-moving mobile world these place-myths are produced, reproduced, contested, developed, circulated, and recirculated through and between the scapes suggested by Appadurai, leading to multiple global and embodied senses of place.

Unpredictable global flows have created images and place-experiences that challenge and subvert the traditional imagery and experiences of Harrogate. With time-space compression the rapid transmission and circulation of imaginary Harrogates is beyond official control, leading to disjunctures not only in image, but in experience for visitors and residents in a town with multiple senses of place, and possible performances of place. There are local and global disjunctures between Harrogate's economy, culture, and politics, with some local imagined communities in particular contesting the development of the 'new' Harrogate. Harrogate's commodified diasporic objects represent both 'traditional' myths of place that are in decline as well as new myths of place, new embodied performances of place within a global era.

Notes

1   To take Taylor's of Harrogate as an example, a simple web search finds numerous sites extolling the qualities and virtues of different blends of Taylor's teas and coffees, from websites for North-American British imports, to tea reviews, to fine food and gifts, to British celebrity chefs, to South African tea. Taylor's tea is the tea used by the English conservation body the National Trust, and the Ritz-Carlton hotel in San Francisco proudly boasts that every afternoon, tea is served that 'includes a perfectly blended pot of fine tea from Taylor's of Harrogate (the official tea of the British Parliament)'. You can buy packs of tea and coffee to take away in a Tokyo hotel, and people cruising in the British Virgin Islands can nip into the port supermarket to buy it.

2   In August 2003 there were 47 listed restaurants in Harrogate, 22 pubs, café-bars, and wine bars, and 15 cafés in a town with a recorded population of just under 73,000.

Chapter 7

Bodies, spirits, and Incas: performing Machu Picchu

Alexandra Arellano

Introduction

No longer is Machu Picchu the best-kept secret place of the Incas, as it was during the European conquest of the sixteenth century. Since its 'discovery' by Hiram Bingham in 1911, the so-called 'lost city of the Incas', located in Cuzco province in the Southern Peruvian Andes, has become a World Heritage Site and a major tourist destination that keeps stirring the imagination of the world (see Arellano 1997). As a result of growing foreign interest in Inca civilization, and for other historical reasons (see Flores Galindo 1987), Peru has branded itself as the 'Land of the Incas' and made the promotion of the tourist industry a priority for the country's economic and cultural development.[1]

The rapid growth of tourism has often driven cultures and places to rethink their uniqueness, identity, and 'brand' by which political authorities package local, regional, and national culture for tourist consumption. However, apart from the cultural and economic consequences of tourism, little attention has been paid to the changing configurations of local places after 'world heritagization' and 'touristification' has catapulted them into the 'global order' (Urry 2002a). Machu Picchu is now part of a large and complex repertoire of images of the media-processed world, where a worldwide audience constructs from different experiences of metropolitan life an 'imagined world' of its own (Appadurai 1996). As a result, the ancient Inca civilization is converted into an object of sporadic dreams, fantasies, and desires of travelling to the Inca 'destination'.

More than dreams or virtual travel, this chapter is about corporeal tourist performances and their role in renewing the semiotic resources of global Machu Picchu through simultaneously producing, consuming, and sensing Machu Picchu. Tourism does not simply involve the development of a hosting infrastructure, and tourists are not only passive gazers. The industry participates in a real mobilization of the imagination and meaning where tourists, as active interpreters *and* performers, significantly imagine and reimagine the contours of the Inca sanctuary. Bodies, spirits, and Incas reflect different forms of sacralization, where Machu Picchu transcends the frontiers of the archaeological remnant and revives the Inca icon into innovative contemporary significances. In other words, analysing how tourist performances contribute to transforming the configurations of the Inca ruins enlightens the contemporary fears, fantasies, and quests of everyday life.

I first discuss the world heritagization of Machu Picchu and how the site has become a theatre of global expressions wherein environmental issues and the threat of global investments are constantly assessed. I then identify two alternative ways of performing and sensing the Incas through the play of highly corporeal and sensuous experiences that engage with the natural, the cultural, and the spiritual. These renewable imaginative processes, incorporated in embodied performances, are intertwined and work as frames of reference for comprehending how the Inca heritage is in play with several global centres.

Saving the 'genuine' Machu Picchu

The inscription of Machu Picchu on the World Heritage List of UNESCO in 1983 led to its formal recognition by the world as an 'amazing urban creation of the Inca Empire' built in an 'extraordinarily beautiful natural setting'.[2] This process of 'world heritagization' impels a process of recognition and authentication that reconfigures the site into a 'global' place and icon (see also chapter 10 on the Taj Mahal). From being a regional heritage that came to be managed, regulated, and protected by the Peruvian state, Machu Picchu is now the object of a common 'global' heritage, a condition that instituted a sense of global responsibility for its preservation and conservation. Machu Picchu became a mirror of humanity, a sample of a 'genuine' and 'authentic' work of nature *and* culture (see also chapter 2 on the Caribbean as 'untouched').

Several authors have located the origins of the heritagization movement in the French Revolution that re-symbolized the 'private treasures' of the monarchy into 'the nation and the people's memory' (see Audrerie *et al.* 1998; Arellano 2000). Could we say that global organizations such as UNESCO are reconfiguring 'national' objects of memory into 'global heritage' by transcending the boundaries of the state and propelling the sites, their images, and their significance into the global order? In 1978, twelve sites were initially listed as being of World Heritage interest by UNESCO. Twenty-five years later, 754 archaeological sites, whole cities, landscapes, natural parks, historical monuments, and buildings had become World Heritage Sites, representing the 'global village' where 176 state members have signed the World Heritage Convention.[3]

This World Heritage 'value' marked by the 'most universal international legal instrument'[4] establishes a worldwide recognition and concern for conservation, but also assembles a place into a newly born tourist attraction (see also chapter 10) involving transnational flows of visitors, capital investments, and an accelerated mobility and visibility on the global stage of the mass media. In other words, the conservation practices imposed by 'world heritagization' contain an ambiguity of protecting a 'genuineness' through global recognition but which also involves alteration caused by the consequent development of the tourist industry. 'Worth conserving' as untouched often means 'worth visiting', and therefore often leads to 'worth investing in', a process involving management and preservation programmes, but also questions of inequalities, contestation, and environmental risk.

'Heritage' and 'tourism' have come to be inevitably intertwined. They are often seen as being in a paradoxical relationship referred to as the 'McDonaldization' process based on a rationalization principle of homogenization that is in opposition to the idea of 'saving the genuine' (Ritzer 1996). For the case of Machu Picchu, tourism allows the world to experience the wonders of the site, and is also of economic and social importance for the region and the country. But at the same time, the industry causes environmental damage and pollution, and attracts controversial investment projects.

President Alberto Fujimori's administration (1990–2001) was one of the first Peruvian governments to rank tourism among the most important national industries. After a decade of terrorist activities, a new political stability and confidence among Peruvians favoured foreign investment, occasioning a steady improvement of the tourist infrastructure. From the second half of the 1990s, international arrivals increased dramatically by 12.9 per cent per year, a substantially higher number than the world average annual growth rate of 4 per cent (WTO). This tendency has also continued until the time of writing under the government of Alejandro Toledo (since 2001), which created the National Tourism Corporation with the goal of attracting three million visitors per year to Peru. In 1992, 94,575 tourists visited Machu Picchu; by 2002 the figure had risen to 457,100 (BADATUR).[5]

From the end of the 1990s, the Fujimori government attempted to privatize sectors of the Peruvian economy, and the tourist industry that had grown up around Machu Picchu was one of the main targets of this drive. Concessions were granted that led to the construction of a cable-car system and the unlimited development of a luxury hotel at the gates of the Inca citadel. The plan also predicted a tourist complex, including a cable-car terminal with tourist boutiques and restaurants. Interestingly, despite the fact that the cable-car project would build a complex 'for' tourism, most of the local and international tour operators, academics, and even tourists themselves opposed the project. As Efrain Bellido, Dean of the Faculty of Tourism from the University San Antonio Abad of Cuzco, states:

> the sanctuary of Machu Picchu is our most precious national treasure, and it is a natural resource that is not renewable. According to UNESCO studies, the building of the cable car would drastically increase the number of visitors to the ruins, from something like 500 to 4,000 people per day, a fact that would destroy the natural and architectural fabric of the complex in less than 10 years. Unfortunately, our government with its neo-liberal politics works in the service of foreign capital and does not take into consideration the devastating consequences. What would happen to our local economy, when English, Chilean, and Spanish investors would administrate and have all the privileges? Moreover, this whole project goes against the very essence of the beauty of Machu Picchu, which is an expression of ancient science, knowledge, and technology, and this cultural contrast with the Inca trails, mountains, and forest would be totally destroyed. The whole nation is opposed to the project, like the rest of the world – look at all the demonstrations here in Cuzco: even tourists were marching with us.
>
> (Interview 2001, my translation)

The world's 'sense of responsibility' rallied to the campaign to 'save Machu Picchu' from becoming 'McPicchu'. At its 23rd session, the World Heritage Centre and the International Council of Monuments and Sites (ICOMOS) undertook a special mission to assess the proposed construction of a cable-car system.[6] As the official website of the movement for 'saving Machu Picchu' emphasized, arguments opposing the project have been carried in media reports around the world. UNESCO strengthened its efforts to hold Peru to its commitment to preserve the World Heritage Site until its full adoption of the UNESCO recommendations of a Master Plan. In May 2001 the project was indefinitely suspended.

The example of this controversy shows the contestation over developing a World Heritage Site where foreign investment, politics, inequalities, and endangered environment issues are constantly generating conflicts among global, local, and national projects. As we can see from the comments of Efrain Bellido, 'local contestation' also entered the global sphere of collective action in order to fight for regional rights. The archaeological complex may have been 'saved' from a cable-car project, but it is still threatened by the chaotic development of the tourist infrastructure, the constant increase in visitor numbers, and the noise and exhaust from the buses that run to and from the ruins.

In general, since the beginning of the new millennium, several new regulations have been implemented in the sanctuary of Machu Picchu and along the Inca trail in hope of preserving the natural and cultural integrity of the park. In January 2001, the entry fee was increased from US$17 to US$50, and no more than 500 people will be allowed to start the trek each day, with a limit of four nights per journey; the trail will be closed one month every year for maintenance; each hiking group will be limited to 30 persons including tourists, cooks, porters, and guides; and all visitors will have to take the trek with a registered company. Moreover, the working conditions of the native porters are now improved, as the weight they carry cannot exceed 25 kilos.

Thus, the world heritagization of Machu Picchu by an organization such as UNESCO, which is one of the most important global 'authorities' in the cultural sphere, is operated through contradictory movements of 'protecting as genuine' but also 'altering for commodification' and developing tourism.

The development of the world's 'sense of responsibility' for protecting the very symbol of Inca heritage is also notable in tourist performances. Alongside the development of mass tourism, tourist performances also follow the 'saving the genuine' trend through by rebranding the site, and trying to find a more 'authentic' way of sensing the Incas (see also chapter 10).

Gazing 'with all the body'

> The Inca trail is the perfect expedition for those who seek to get in tune with their inner Indiana Jones.[7]

On the margins of 'mass tourism' day-return trips to the ruins, the self-distanced and conscious 'post-tourists' (Urry 2002a) aim to perform practices of conservation, at the same time as supporting causes such as the protection of endangered sites. 'Travellers' or the 'un-tourists' (see Birkett 2002) are aware that tourism has to be compatible with preservation, and many of them follow rules to preserve their 'alternative' way of experiencing Machu Picchu (see chapter 8 for a discussion on 'experience'). The four-day 'Inca trail' leading to the citadel is promoted as 'ecotourism' for a responsible 'ecotraveller' (see also Duffy in chapter 4). Several new lodges, hotels, and travel agencies in Cuzco and Machu Picchu now emphasize their 'eco-operated ecotechnologies'.

Tourists are now 'environmentally aware' and opt for ecotravelling, a market set to flourish in the coming years.

The rebranded 'ecotraveller' differentiates herself from the (mass) 'tourist' who is constantly being 'condemned as anti-ethical' (Birkett 2002). But more than promoting ecological performances based on sustainable tourism programmes that minimize impacts on a pristine area, ecotourism is here conceptualized as a more 'authentic' way of experiencing through fully corporeal practices that challenge the body and seek a full and revitalizing contact with nature (see chapter 4 for definitions of 'ecotourism').

Do we all have an 'inner Indiana Jones' or something like a sense or desire for adventure? During the 1980s, a growing interest in hiking the 'Inca trail' helped transform the citadel into a global place for adventure. The trek is imagined and performed as restoring and resensitizing a so-called 'metropolitan body' into a highly sensuous corporeal body (see chapter 5). In his discussion on Simmel's perception of the 'modern individual', Lewis describes an 'irredeemable rupture for embodied experience', where 'there is an intensification of nervous stimulation' as 'our senses become besieged by the information overload endemic to modern life' (2001: 65–66). Therefore, in the world of the adventurer, automatic and passive metropolitan bodies are in search of enjoyment and at least a momentary recovery of fully corporeal and highly sensuous activities (see chapter 5 on the 'moment' in surfing).

From the ecotourist viewpoint, a tourist who does not enter Machu Picchu by trekking the Inca trail has not properly experienced or felt what it was like to live in the Inca era. The four-day trek is essential in reaching this sensuous state, as the long-distance walk and camping among nature, beauty, and ancient civilization allows the achievement of a reflexive awareness of the body and senses, and therefore a fully corporeal experience of the sacred site. Since the early 1990s, the Inca trail has become a tourist rite of passage that testifies to the 'real Inca experience'. To gaze at Machu Picchu, 'to be there', to take a picture of oneself in front of the temple of the sun as a proof that 'I was there', is not enough. The visit requires sacrifice, challenge, physical endurance, and a wide range of multidimensional embodied performances that convert 'touring' into 'performing' and give way to a self-transforming experience.

The main physical activity involved in performing the Inca trail is obviously walking. But how has 'four days of gruelling hiking in various conditions' – which is rather different from walking as a leisure activity – become so attractive for travellers? Why does the tourist seek physical challenge, sacrifice, pain, tiredness, illness, *soroche* (altitude sickness), cold, stomach problems, sunburn, and so on? Improving the over-socialized self by stretching one's physical capacities and reaching the confines of sensuous experience are plausible answers. As Edensor comments, 'beyond the trial of physical endurance and mental strength lies the promise of a more confident self and a return to a masculine (bodily) essence, replete with fantasies about getting back in touch with (one's) nature' (2001b: 93).

The hike starts at an altitude of 3,000 metres (9,840 feet), ascends to 4,200 metres (13,780 feet) at its highest point and ends at 2,300 metres (7,540 feet) when reaching Machu Picchu. The second day of the trek is the most physically demanding, and many people have difficulty in reaching *Warmihuanusca* (Dead Woman Pass), the summit of the trail. This pass is reached after 3–5 hours climbing a mountain with a temperature that ranges from the semi-tropical climate of the rain forest to below freezing. Indeed, the whole route is tough, stony, and composed of many steep ascents and descents. The idea of carrying a heavy backpack, the fear in general, the blisters, the flies, and other obstacles make the trek a physical challenge that has to be accomplished successfully. Walking here is not simply the embodied practice usually seen as central to tourism. It is rather an arduous journey made on foot, which demands a sense of adventure. Jen Cravens, a young traveller, writes on how she 'survived the Inca trail':

> I was terrified of the hike, of the pain and strain and unknown obstacles. But I had come this far, and I would never forgive myself if I allowed fear to dictate my actions. So with bated breath and a knot of dread in my stomach, I agreed with my companion, William, that I would choose the adventure of the trail over the ease of the train. With a heart filled equally with hope and fear, I left Cuzco…where we began our adventure.[8]

The Inca trail requires physical and mental preparation, and most people who brave it are not passionate or experienced hikers. In fact, the more expert hikers avoid the trail during the high season when novices seem to be 'queuing up' in the mountains. Therefore, the trail is a challenging rite of passage and a unique experience that requires physical and mental conditioning.

> It was a bit risky for me because I had an injury in my knee. I needed a good bandage and intensive training in the Gym three weeks before travelling from Chile. My boyfriend also agreed to carry my backpack…I was scared but I really wanted to do it. Suffering while being filled with wonder can't be more thrilling.
> (Ana, Chilean worker, interview, my translation)

> I didn't prepare myself physically but after the second day I thought it was rather easy, so I was walking very fast on the most difficult day! I am so stupid because on the third day I couldn't breathe from pain in my sore muscles, my boyfriend had to carry me!
> (Camille, French student, interview, my translation)

Tourists hike towards Machu Picchu by confronting their bodily limits and undertaking a battle against nature. But in addition to walking, there is another interesting and surprising way of 'performing' the Inca trail. A Peruvian native now living in Los Angeles, Devy, defines himself as a 'multiday ultra wilderness running freak' who has been guiding runners in the Andes for the past three years. During a trip to Cuzco in 1995, Devy decided to 'run' the Inca trail in a single day. This experience gave birth to an increasingly popular 'tour' he now leads several times a year. Running the Inca trail is a much more formidable challenge than hiking it. Confronting wilderness, altitude, and the unknown is the best way to 'gaze with the whole body'.

Many people also regard taking the train to Machu Picchu as a great experience, although walking and camping for four days on the Inca trail to finally reach the magnificent view is certainly more taxing and more valued on the modern scale of achievements. However, *running* the Inca trail (10 hours 43 minutes is the current record) instead of covering it in a four-day walk is an exploit that can hardly be surpassed, and testifies to the modern need for some kind of extreme physical challenge.

As one runner put it: 'I was turning 50. I had seen so many lives turned upside down by the "mid-life crisis" thing that I decided I'd better do something to hit it head on to be sure the "fall-out" wasn't too devastating. Thus Peru.'[9] As Edensor stresses, 'fulfilment takes the rhetoric of individual achievement counterposed to the regulations and fetters of everyday family and work life, connected to a "can do" philosophy of personal growth and reflection' (Edensor 2001b: 95). From 'I was there', to 'I did it [the Inca trail]' and from 'I hiked there' to 'I ran there' indicates the increasing importance of claiming status through physical achievement where pleasure and competition, athletic performance and touring are new and flourishing ways of sensing the Incas. These experiences push the body to extremes and achieve a reflexive awareness of the body.

A site with such qualities is becoming appealing for world travellers and for people who cherish everything related to 'nature'. One issue that Macnaghten and Urry (1998 2001) have explored is the emergence of a 'culture of nature' in most Western countries. According to them, this phenomenon is 'a culture that emphasizes valuing the natural, purchasing natural products, employing images of nature in marketing, supporting organizations concerned with conserving nature, being in the natural environment and engaging in practices that enhance the "naturalness of one's body"' (Macnaghten and Urry 2001: 1). In the adventurer's imagination, it could be said that beauty is evaluated in terms of environment, ecology, fauna and landscape in relation to the sensuous contact of the body-subject with nature. The Inca trail to Machu Picchu reflects the very performance of 'sensing' the Incas, as the traveller experiences a grand vision of nature and the mystery of the past.

The Incas are always depicted as a people who were very close to nature, and Machu Picchu is a monumental demonstration of that intimacy, with its terraces and buildings constructed in perfect harmony with nature and the mountains. As every tourist knows, the Inca praised the Sun, the mountains, *pachamama* (Mother Earth) and all forces of nature. As they had neither horses nor wheeled transport, the Incas relied on walking. Therefore, entering Machu Picchu through the Inca trail is a way of experiencing the Inca nature by approaching the city as they did themselves – reviving corporeal past performances. Jen Cravens chose to trek the Inca trail in order to 'enter the city the way the original inhabitants did. It was more important to me that my steps echoed the Incas.'[10] While gazing with all the body, the adventurer is alternately in 'play' with the natural and the cultural, through corporeal performances that are aimed at being more 'real' and authentic than those of the 'passive gazer' of the packaged tour. Behind the choice of a challenging bodily experience over the ease of the package tour lies a search for authenticity based on a 'sense of reality' that is achieved through re-enacting the Incas' past performances. Walking, being close to nature, and recovering momentarily the fully corporeal and sensuous experience of an ancient civilization requires a sense of adventure and of sacrifice that braves bodily discomfort. The desire to exceed the 'passive' gaze of the 'archaeological site' by engaging the whole body and senses adds to the experience of touring and transforms the sanctuary into a theatre of re-enacted Inca civilization wherein actors perform the past and get closer to the 'real Incaness'.

Quest for the 'sacred site'

Another possible way of performing Machu Picchu is often expressed by Cuzco inhabitants when they refer to what they call 'esoteric tourists'. Even if these performances seem to have always existed on the margins of an 'institutionalized' tourism, these spiritual seekers are increasingly visible in the region, and local people involved in developing 'mystic tours' are doing very good business as a consequence. This 'alternative' way of experiencing Machu Picchu and the region of Cuzco is expanding as several 'mystical journeys' and 'healing conferences' invite tourists to recover lost wisdom and accelerate spiritual growth. The mystic tourist clientele, typically middle class and from historically Protestant countries, is likely to spend more than average tourists in order to experience a transformational journey. Today, several international tour operators promote Machu Picchu as a 'power place', and bring thousands of visitors to get in touch with their 'creative energies', to return to the basics and reconnect with the sustaining power of the ancient civilization.

Several authors have associated this phenomenon with the New Age movement, which is seen as the continuation of the counter-culture movement of the late 1960s (see Heelas 1996; Bloom 1991; Bellah 1991; Hetherington 2000). 'Meditation, shamanic activities, wilderness events, spiritual therapies and forms of positive thinking' are all key ideas and practices related to the movement (Heelas 1996: 1). We saw that adventurers or ecotravellers construct their performances on a differentiation from 'mass tourists'. Here, 'New Age' travellers seem to reject most of the official performances and discourses of the Inca citadel. From the 'mystic gaze', culture and nature are combined in the 'sacred' where Machu Picchu is more than a place to play; it is a 'power place' where the 'trainee' performs a self-declared 'quest' in order to receive a local teaching. Machu Picchu is not only a 'symbolic' global place, it is a therapeutic centre that provides the required energy and indigenous spiritualities for reaching a 'liminal' state (Turner 1969) or a transformational experience with the 'beyond'. These 'self-authoring' practices elevate Machu Picchu into a global network of 'power places' where ancient and prehistoric civilizations 'join up' by means of linked 'energy spots' such as Stonehenge, Egypt's pyramids, Mount Everest, and 'mystical temples' such as the Taj Mahal, forming a 'global' pagan path.

Machu Picchu is located amidst a powerful natural landscape where the civilization of the Incas is seen as the 'spiritual heritage' of a people who 'worshipped' nature, stars, and healing energy. As is well known, the Incas described themselves as 'Children of the Sun' and in their mythology they were descended from celestial realms. The fact that popular ideas claim that Machu Picchu was an Incan 'astrological centre' or 'astronomical observatory' have also led to claims by New Agers that the Incas were in contact with extraterrestrial life. The attempt to find some hidden 'meaning' in the architectural geometry of the citadel (such as the temple of the Sun and its 'solar clock' that measures time according to the position of the Sun's shadow), and in its location, height, and position in the middle of the Andes, are some of the features transforming the Inca archaeological site into a 'sacred sanctuary' and one of the greatest 'power places' of the world. In this context, the fact that the Incas had no written records – and consequently that no detailed information is available on their beliefs, knowledge, or motivations – creates a convenient blank space in our knowledge about them onto which competing New Age interpretations can be projected without fear of any definitive contradiction.

The 'mystic gaze' can be undertaken by a series of performances and quests that are not always easy to grasp. The 'spiritual guides' working with tourists are often local shamans who teach and 'inspirit' the 'disciples' towards an awareness and control of their inner powers, which they claim to be the Inca way. In a mode of promoting tours that might not be out of place advertising a conference venue, New Age tours declare that 'many speakers offer opportunities for one-to-one consultation that can be truly transformational'.[11] Touring, learning, and improving oneself and one's relationships with others by learning how to 'magnetize energies' are among the basic ideas that motivate the quest.

These New Agers perform not only in uncommon grounds but also in prohibited spaces to avoid crowds and experience the very essence of the 'energy' they are seeking. Entering Machu Picchu by night, for example, is to take advantage of solitude, silence, starlight, and sunrise. Aluna Joy Yaxk'in, a spiritual guide who leads and organizes several pilgrimages to Machu Picchu and has travelled to several sacred sites, prefers to enter these power places at night to meditate and pray. A night of meditation in these sites has proven 'a profound way to tune in and receive sacred mystery school teaching':

> On my last day in Machu Picchu I was ready to enter the site at night. I waited until 1 a.m. to make sure the guards were asleep. I climbed over a cliff beside the ticket gate to enter the site. I ascended part of the Inca trail to a special carved rock where I could see the entire site laid out before me. Machu Picchu was exquisite, lit up by starlight, and the silence was profound.[12]

Spiritual pilgrims experience Machu Picchu differently from the adventurers or the regular day-return tourists. This alternative way of gazing with the whole body and spirit favours the 'learning experience' – at least, for those predisposed to believe there is a spiritual lesson there to be learnt. In Aluna Joy Yaxk'in's words: 'we do not learn solely with our minds, as we have been taught in the modern world. We learn with our entire being, and our entire being will shift because of this learning. We understand that the living library locked within our DNA is in every single cell of our body. This is whole body activation and is exactly as the ancients intended us to learn.'[13] Here, the use of the body is different from that of the ecotourists, as the 'New Agers' do not usually drive their bodies to extreme physical challenges, although they still seek a certain 'embodied sacrifice'. 'Enlightenment' and the 'liminal' phase do not come after a simple 'unusual' gaze like 'gazing at night'; there is, it is claimed, hard work to do to reach the 'anti-structure' (Turner 1969) that uses the body and the senses differently. In fact, after meditating, freezing, and praying through the night, Aluna expected a 'teaching' or 'divine message'.

> I sat down with my back against the rock and began to meditate. The stone was getting colder and colder and it was getting harder every moment to concentrate. I...could hardly move but the cold intruded into my meditation with a vengeance. I sat all night in the cold. By morning's first light I had not received a single vision, insight or teaching...As the sun rose over the top of the surrounding Andean mountains, I greeted the warmth it brought. The vision of the sunlight hitting the ruins below was an amazing view and a seeming consolation for my night of suffering.[14]

New Agers seek personal enlightenment, teachings manifested by signs from 'another dimension'. They see the Inca ruins in the middle of nature but they 'force' the gaze in order to detect particular lights, stars, angel shapes, and any sign that, they believe, comes from 'the beyond'. New Age travellers see themselves as 'students' who learn how to see these mystical signs and how to interpret them properly in order to understand their teachings.

While facing a glorious sunrise at Machu Picchu – without 'mystical vision' – Aluna waited for the right moment to snap her first photograph. At the same moment, a couple of tourists appeared right in the line of the photo emerging in front of her. It was 6 a.m., and the doors were not supposed to open until 8. 'Where did these bozos come from?' she wondered. The fellows then pulled some bowling pins out of a bag and began to juggle. 'It was the most absurd sight I have ever witnessed after an all-nighter in a sacred site…It was as if God was playing a big joke on me.' Years passed and one day, one of her 'tourist-students', after long hours of meditation, told Aluna that he received a message from a 'Lumerian Master' saying that the key to enter the 'Golden City' was to juggle. As she writes, 'I have been calling the Masters to come to me in physical form all my life, and here they were right in front of me, and I did not recognize them!'

How to read 'signs', how to hear 'messages', how to sacrifice the body in order to feel the 'beyond', and how to look at the landscape in order to transgress into 'another dimension', are different and increasingly popular ways of sensing Machu Picchu and its mysterious power. Through meditation, Aluna and New Agers are looking for enlightening 'visions' where the gaze seems to be primordial but at the same time it is intended to be subordinated to the authority of the self. Here, the 'turn within', or what Bellah (1991) called the 'internalization of authority', is motivated by the loss of faith in the authority of institutions. In other words, the self becomes the only truth, and the Incas act as models, ideal healers, and 'voices' towards an authority that has not been 'contaminated' by the institutionalization that is imposed on things such as family, education, and work by modern society.

Modernity generates uncertainties, which some attempt to overcome by quests for an 'authentic experience as a means of revitalizing fragmented personal identity' (Heelas 1996: 148). Going further than MacCannell's search for authentic 'untouched' cultures (1976), the New Agers locate their 'quest' in complex spatialities (see chapter 14 on authenticity/spatialities) that do not simply gaze at 'indigenous authenticity' but are constantly in 'play' at different levels, engaging the body, the senses, and spiritualities with, as they would have it, the Incas, the stars, and the beyond. The New Age 'experience' is thus aesthetic, kinaesthetic, and also self-transformational and 'healing'.

The mystic gaze seeks a reality located inside the self rather than outside, and which uses the mystical 'vision' as a medium to reach the inner self. At the same time, it rejects the 'socialized gaze' and 'institutionalized' self. Indeed, in the New Age philosophy, the senses are to be avoided in order to reach the path to interiority. As Bloom puts it, their belief is that 'all life as we perceive it with the five human senses or with scientific instruments, is only the outer veil of an invisible, inner and causal reality' (1991: 33). Aluna's meditation consisted of forgetting about the cold stone, the cold, and exhaustion. Transcending the body, and rejecting the senses as being 'infected' by the pre-given order of things, is, in her view, the path towards the authenticity of the 'within'.

Bellah (1991) goes even further, saying that the quest is 'beyond belief'. In his view, therefore, healers, shamans, New Age teachers or any charismatic leaders should not impose anything upon their 'students', but 'guide' them towards their own 'soul', in the process using 'inner voices' and intuitions to force out the student's 'false person-ality'. Paraphrasing Heelas (1996), Machu Picchu is the stage and the context that enables travellers to reach their own spirituality and authority.

Conclusion

Adventurers and New Agers aim to differentiate themselves from mass tourism prac-tices that are often judged as being less 'authentic', more superficial, and anti-environmental. After becoming a World Heritage Site, to be protected as 'untouched', performances at Machu Picchu have been rebranded as 'life-changing opportunities for you and the environment' (Birkett 2002). Self-sacrifice and self-transformation through meaningful experiences aim at improving the self – spiritually and bodily – at the same time as performing the action in an environmentally respon-sible way. Accordingly, one could say that these rebranded performances, based on a 'do-good' philosophy (for the site and for the self), intend to veil the ambiguity of conserving *and* commodifying heritage sites, as travellers 'buy' more 'authentic', 'responsible', and 'alternative' performances that are in fact still unalterably staged. Thus, after the world heritagization of Machu Picchu, bodies, spirits, and Incas are now haunting the reassembly of the 'secret place' of the Incas into a playful global place for contemporary urban quests.

Notes

1 This paper is based on a chapter of my doctoral thesis, carried out in the Department of Sociology at Lancaster University. I conducted interviews, participant observation, and research in travelogue websites (see Arellano 2003).
2 *http://whc.unesco.org/nwhc/pages/sites/main.htm*
3 *http://whc.unesco.org/nwhc/pages/doc/main.htm*
4 *ibid.*
5 'Banco de datos de turismo', University San Martin de Porres, Lima, see *http://badatur.turismo.usmp.edu.pe*
6 *http://whc.unesco.org/archive/repcom98.htm#sc274*
7 *www.andesadventure.com*
8 *www.jencravens.com/*
9 *www.runningclubnorth.org/peru_1.htm*
10 *www.jencravens.com/*
11 *www.powerplaces.com*
12 *www.kachina.net/~alunajoy/2001jan.html*
13 *www.kachina.net/~alunajoy/99april.html*
14 *www.kachina.net/~alunajoy/2001jan.html*

Chapter 8

On the track of the Vikings[1]

Jørgen Ole Bærenholdt
and Michael Haldrup

Here terrible portents came about over the land from Northumbria, and miserably frightened the people: these were immense flashes of lightning, and fiery dragons were seen flying in the air. A great famine immediately followed these signs; and a little after that in the same year on 8 January the raiding of heathen men miserably devastated God's church in Lindisfarne island by loathing and slaughter.

(*Anglo-Saxon Chronicle*, entry for the year 793 (2000: 15))

Michael, wouldn't you like to be a Viking, sailing all over and meeting a lot of women?
(Overheard at the Viking Ship Museum, Roskilde, July 2003)

Introduction

Few icons, images, and archaeological antiquities are as widespread in the North Atlantic region as those relating to the world of the Vikings. The Viking past is a major topic in contemporary national education, in historical and archaeological research, and in many museums. However, it is not confined to these spheres, but appears across a vast spectrum of cultural fields such as politics, commercials, cartoons, action movies, religious beliefs, and role-playing games. Many clubs and societies have been established to conduct experimental investigations of Viking food, textiles, crafts, shipbuilding, religious cults, and so forth. Traces of Viking culture can be found in places throughout Europe (including Russia and Istanbul – the former Constantinople), across the Atlantic in Greenland and North America, and even farther afield, since recent research indicates that there were Viking trading posts in Baghdad and North Africa. Viking museums, sites, and replicas are spread throughout the North Atlantic region. Heritage industries offer re-enact-ments of Viking warrior and craft skills in Newfoundland, Greenland, Iceland, the Faroes, Ireland, Great Britain, Norway, Sweden, and Denmark. In North America tourists can visit the first confirmed European landfall in the Western Hemisphere, the L'Anse aux Meadows National Historic Site in Newfoundland. They can take a 'time trip' back to the age of the Vikings at the Jorvik Viking Centre in York (Addyman 1994). And in Southern Greenland they can visit 'Thorhildur's Church', reconstructed for the events (including staged Viking–Inuit fights) in 2000 that commemorated Leif Eriksson's journey from his Greenland base to set up the North American colony of Vinland (Bærenholdt 2001). In Britain, Ireland, and Norway, tourists can ride on Viking roller-coasters and go on boat trips such as the 'Viking Splash Tour' in Dublin where people can take a ride through the town in an amphibious vessel while wearing horned helmets. In other places, such as the site discussed in this chapter, attempts are made to re-enact the practice of 'authentic' Viking skills, for example house construction and shipbuilding, for tourists and other visitors. While the purposes of such Viking re-enactments vary from explicit experimental research to pure amusement, the genres increasingly interact, making the world of the Vikings seem omnipresent in commercials, popular culture, heritage industries, museums, and in national, regional, and social identity projects, and so on.

Though the Vikings have been dead for nearly a thousand years they still seem to haunt the living. A recent BBC-initiated project to use DNA analysis to track the 'blood of the Vikings' in the veins of contemporary inhabitants of the British Isles (in the TV series and book of that title hosted and written by the archaeologist Julian Richards) is a sign of the fascination with Viking heritage and genealogy that reaches beyond the borders of the Norsemen's historical homelands (see Richards 2001). To postmodern Europeans and their descendants, then, the Viking past is far from remote, but is an ever-present source of fascination, interest, and imagination. In this chapter we want to explore this by asking what power of fascination directs people's imaginations and travel itineraries to the Viking sites and makes these seem ever present and omnipresent. Is it a quest for origins, edification, and authenticity? Is it simply a matter of amusement and entertainment? Or do we have to address more complex issues about how identity-formation and pleasure-seeking intersect?

This chapter is based on a short field study at the Viking Ship Museum in Roskilde, Denmark, in July 2003.[2] The museum contains a unique exhibition of five well-preserved Viking wrecks excavated from Roskilde Fjord in 1962, and is firmly rooted in the tradition of museums of cultural history, with an emphasis on its specialized knowledge and skills in experimental maritime archaeology and the construction and sailing of replica ships. Since 1997 the museum has increased activities that involve visitors more directly, primarily through the construction of a 'Museum Island' for more activating and staged activities related to the world of the Vikings (figure 8.1). Thus the Viking Ship Museum displays some important contradictions and ambiguities relating to how heritage is presented and interpreted in contemporary tourism.

Debates on heritage tourism in general and Viking heritage in particular have revolved around the concepts of 'authenticity' and 'commoditization' (MacCannell 1989; Cohen 1979, 1988; Redfoot 1984; Wang 1999; Halewood and Hannam 2001; Olsen 2002). Much of this debate has falsely presented the question as one of either/or, presenting museums as necessarily being *either* serious *or* entertaining, thus neglecting the ways in which visitors ascribe meaning to and inscribe their own practices within heritage sites. In this chapter we shift perspective from the barren objects and sites of heritage tourism to the practices and performances of tourists. In doing so we approach the Viking heritage as 'sites of interpretation' (Crang 1994, 1996). While the first section of this paper explores how Viking heritage is encountered and interpreted by tourists visiting the Viking Ship Museum, the second section suggests a framework for understanding how material encounters and imaginative fantasies intersect in the performance of heritage places. In the third section we discuss the transformation of Viking heritage from national stereotypes to global traces and trails.

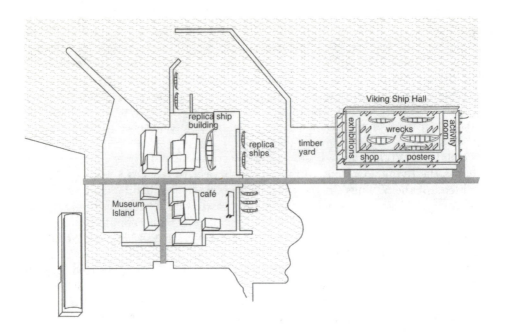

Figure 8.1   Viking Ship Museum map

Figure 8.2   The wrecks, Viking Ship Hall

Figure 8.3   Activity Room, Viking Ship Hall

Figure 8.4   Construction of a replica Viking ship

Reading Viking traces

To most of its visitors, the main attractions of the Viking Ship Museum in Roskilde are its skills and knowledge in traditional shipbuilding. Although the remains of the five Viking wrecks are the central objects exhibited in the 'Viking Ship Hall' (built in 1969; see figure 8.2), they are mainly gazed at in passing while visitors move through to view other parts of the museum. The typical flow begins with the 'Vikings in Ireland' exhibition and the IT-based exhibition about building replicas, entitled 'From Wreck to Reality'. After passing the wrecks, visitors enter the hands-on Activity Room (see figure 8.3). Here they can enter two replicas of parts of ships, touch Viking cloth, food, furs, ropes (all replicas), dress up as a Viking with weapons, write in runes and play Viking games, all carefully observed and accompanied by actively communicating students. Finally, visitors can stop at posters, for example a display of Viking sailing routes, before arriving at the museum shop. Outside the Viking Ship Hall there is a small open-air museum surrounded by water channels: the Museum Island. A collection of traditional Nordic ships (replicas) is anchored here. A ship is being built in the open using original techniques (figure 8.4), and tourists can participate in sailing tours, knitting, painting shields, making coins, or perhaps visit the archaeological workshop, or simply have a pleasant time with refreshments in the central café area. There are also a timber yard and specific projects for limited periods such as the re-enactment project called 'The Warrior Practices'.

The self-perception and image of the Viking Ship Museum is closely tied up with research and skills in producing and sailing replica ships, and with the knowledge of researchers, craftsmen, sailors, and guides (students). But to visitors, it is the interplay between the search for origins and the fascination with the skills, characters, and adventures of the Vikings that frames their perception of the exhibitions. In this section we will examine how the multiple activities and objects at the museum are framed and interpreted by visitors.

The tracing of genealogy (physiological or cultural) seems to be considered particularly important by Nordic visitors such as the Icelandic couple (interview 13) who, after attending the Copenhagen Jazz festival, decided to visit Roskilde, a place well known from their school classes and books. To this couple the cultural heritage of the Vikings is self-evident: 'We have very special names, my father['s name] is Viking, and her father is Odin'. Naming is crucial in Icelandic heritage, which retains a patronymic system in which children take their father's given name as their surname. Names are thus literal descriptions of a person as well as a designation, giving the name a more direct genealogical meaning and identity than in cultures that use settled surnames. The male respondent *is* Christian Vikingson (Christian, Viking's son) and the woman *is* Gudrun Odinnsdottir (Gudrin, Odin's daughter). This genealogy of names also points to their hybrid Christian/pagan heritage, a heritage also handed down by the Icelandic sagas: 'of course, we know about the sagas, and we learn a lot about them in school, and maybe it [the experience at the museum] is seeing them for real', and they talk about excavations close to their home town. Another Icelandic family (interview 10) explained the significance of the Vikings to them in terms of 'beautiful stories'. To Nordic visitors the Vikings, their history, relics, and skills are self-evident parts of their own past (as they are of the authors').

Though Viking iconography and history have played significant roles in the formation of national cultures in the Nordic countries – not least in Iceland, Norway, and

Denmark – they carry connotations that reach beyond current borders. The 'beautiful stories' of a common past may even extend beyond what are normally regarded as Nordic regions, as when a French visitor claims that the museum experience 'is about the European past' (interview 6). While this tracing of cultural roots is indeed an important theme among visitors, the fantastic world constructed by popular culture is equally important to tourist interpretation of the Viking relics and re-enacted skills displayed. Two young sisters from New York and California (interview 11) who grew up in Taiwan, stayed in Copenhagen for one week. Their visit was the result of a coincidental combination of circumstances: their choice of Denmark as a destination for its 'mild weather', and of the Viking Ship Museum from a sightseeing tour programme they consulted to find out 'what to do' on their own. They had little advance knowledge of the Vikings and their geography apart from history classes and cartoons depicting Viking warriors with 'horned hats'. In response to the interviewer's second question about Viking images one sister answered: 'I think to me, though, that it is [the image of] an invader; they invaded – for a period of time they dominated a lot of countries, and it is more of a negative image'. The interviewer's intervention about Vikings in Newfoundland reminds the other sister of something:

> Oh, actually, I watched a TV programme on the History Channel talking about that, and they were trying to figure out who actually discovered America first – Columbus or someone else – and they did mention Vikings, but at some time they mentioned the Chinese sailors, I don't know how many years ago, and the whole programme struck me very much, because of the Chinese, not because of the Vikings...we grew up in Taiwan...so I hoped the Chinese discovered America.

They are eagerly positive about their experiences at the museum and all the accounts given by the guides and the film, fascinated with a world totally detached from their own and their particular heritage. This fantastic world of the Vikings is particularly important in triggering off the hands-on activities in the rather dark, adventurous Activity Room, where one can dress up as a Viking in coats of mail and handle swords, play games, and write in runes. Visitors on sightseeing tours from Copenhagen put a lot of effort into their activities, including photography, when they arrive at this rather dark room to dress up playfully after passing through the simple, rigorously modern exhibition of the original remains of the ships. The attitude of the American-Taiwanese visitors above to the objects displayed is also observable among sightseeing tourists on organized tourist trips (Larsen 2003: 151ff).

However, the fantastic elements of the Viking world also play a significant role for visitors trying to trace ancestors or cultural heritage. A middle-aged couple from Hawaii were on their way through Denmark, the Frisian Islands, and the Netherlands (interview 1). They had visited the Museum Island the evening before and had come back on this bright summer morning (the Museum Island is open to the public) before the museum opened. As they put it, they were 'looking for dead Europeans' – seventeenth-century ancestors from Norway–Denmark and the Frisian Islands. Standing beside the 30-metre-long ship replica under construction, they were fascinated by the 'lifeworld' of the Vikings as directly compared with their own environment and modernity:

She: For us it is very hard to understand living on the ocean, living on a ship for an extended time, because we have seen the ocean where we live [Hawaii]; sometimes it is very calm, but at other times, like during hurricanes, it is very wild and…they were very strong people, very creative and very clever at figuring out how to survive.

He: …and it's unbelievable that they could build anything this modern.

Vikings are admired for being 'creative' and 'clever' and for having built and sailed ships that were 'this modern'. This belief in the fantastic world of the Vikings is more than just praise of their skills. Skills are associated with characters and in turn they relate to their own genealogy by joking about the Viking-like characters of their relatives (fishermen of Norwegian origin in Seattle). The very reason for going back this morning was to take photographs of the replica under construction, 'because we are trying to explain to our grandchildren – their heritage, and this helps…And I'm so interested and amazed that long ago people had developed crafts; they were not literate, they had to make it originally'. Indeed, the genealogical search is a performance of heritage over long distances, where photographs help to build the connections and identifications. These identifications are performed as part of the general ancestor-search programme of the European tour, where they compare the geographies of challenging seas, personal characters, and the building of skills across time and space.

Another American couple, from Utah (interview 3), talked about connections of the same kind. Their grandparents were among the many late-nineteenth-century emigrants from Denmark to America. Again, genealogy was associated with the Viking world:

She: We've heard about Vikings all of our lives – in history we studied in school, and of course being Danish I was interested in anything connected with the Danes.

He: … and the Vikings, as I understand it, were very strong, outdoor-type people. Today we live in nice houses, have nice things – the Vikings didn't have that, they had a strong character, different than people today, different than American people, Danish people, it was a different culture altogether…heritage, it helps us that our *forældre* [the Danish word for 'parents' – he may have meant *forfædre*, 'ancestors/forefathers'] were good people; they were trying to live the best they could, at the time when they lived, and they liked adventure…

She: They weren't afraid to try new things and tried to make their lives better, you know; this is the way I feel about my forefathers when they came to America…

The Viking world is a world of adventure, open to innovation; and interestingly, these capabilities are directly associated with the interpretation of ancestors' migrations to the deserts of Utah, a place that was hard to understand for them compared with 'green Denmark'. While the ships evidently triggered off their imagination, the central concern of this couple was to resurrect the Vikings, their morals and 'life-worlds' as role models for present-day beliefs and ways of life.

Fantastic realism

The Viking Ship Museum obviously affords a wide range of interpretive possibilities. People engage in different practices, and there is an equally wide range of options for linking up with other aspects of the world of the Vikings. However, there is always a dynamic interplay between an interest in the real object and the fantastic iconography that people carry with them from other media. The fantastic world of their imaginations goes hand-in-hand with the tracing of their own pasts, and has significant implications for how they interpret the re-enacted activities and skills, as well as the original Viking ships. The free play of fantasy and the search for authentic knowledge about their ancestors and their cultural heritage intersect and blend in important ways. We therefore dwell here on the workings of fantasy and experience in heritage interpretation.

The hegemony of a certain jargon of authenticity in writings on heritage tourism has had a pervasive and negative effect on interpretations of heritage development. Not only has it refrained from engaging with the contingent circumstances that produce heritage sites such as museums (Meethan 2001: 112); it also applies a shallow, conflated concept of experience. As Negt and Kluge argue, the 'experience' of heritage tourism should not be thought of in the sense of experiencing a bodily sensation, but rather (in the sense of the German *Erfahrung* rather than *Erlebnis*, both of which can be translated as 'experience'), is best understood, they argue, as multi-layered processes, each following its own time-rhythm according to the objects with which the body engages (Negt and Kluge 1993: 18ff). In this process the 'workings of fantasy', a fundamental human capability, according to Negt and Kluge (1987: 60ff), are what enable subjects to contextualize, anticipate, and draft worlds out of an inconsistent mess of drives and experiences. Hence, fantasy is intrinsic to any production of authentic experience (Negt and Kluge 1993: 32ff). The point of Negt and Kluge's approach is the simultaneous interest in real material objects *and* in fantasies. It is a way of understanding how bodily encounters with objects and imaginative practices intersect. Experience (*Er-fahrung*) is more than a question of triggering events of the emphatic kind. People learn over time (*Er-fahren*), as they *fahren* – travel – move through space. People moving through a museum familiarize themselves not only with different objects, displays, and performances, but also with a world of a 'second order', a world of imaginings and fantastic connotations that frame what they see and encounter. As we have seen, this does not necessarily mean they drift away from 'the real', but in important ways it facilitates the interpretation and reading of the past.

The traditional paradigm within museum and heritage displays has to a large extent relied on a 'hegemony of the eye' that separates the auratic object from the gazing spectator (Hetherington 1999, 2002). Furthermore, museum displays have constructed particular epistemic regimes by determining how the world and its past are ordered, classified, and framed (Hooper-Greenhill 1992). But the rise of the heritage industry has in important ways challenged the authority of the museum (Walsh 1992). Particularly provoking to traditional museums such as the Viking Ship Museum is the possibility that untrammelled fantasies of a spectacular past may supersede the aura of the real 'authentic' objects. Visitors often indicated that they read re-enactment activities and replicas as being more real than the wrecks displayed in the hall: 'it does say much more to see them for real, in the harbour – and these [wrecks in the Hall]. If I've seen one of them, then it's enough; four [there are actually five] is far too much, when you have the ships outside' (interview 7, see also interview 13 above). This conception of 'the real' only has meaning insofar as one understands reality not as equal to the objects of the primary world but as a fantasized 'second-world' resurrection of how it 'really' was. In contrast to the displayed wrecks (not to be touched) the replicas could be used and tried out on the fjord. Hence they could provide a corporeal sense of how it 'really' was to be a Viking at sea, in combat and so forth. This is a corporeal reality that also frames the encounter with the five wrecks in the Hall.

In his discussion of the authenticity of art objects, Benjamin distinguishes between auratic objects and 'traces': 'The trace is the manifestation of a closeness, however distanced it may be. The aura is the manifestation of a distance however close it may be. In the trace we enter into the possession of the thing, in the aura the thing over-powers us' (Benjamin, quoted in Markus 2001). These visitors do not gaze at the auratic objects on display, but read them as traces. The wrecks of the tenth- and eleventh-century ships on display in the museum are certainly framed and exhibited as auratic objects. However, it is the replicas, the re-enactments and so forth that enable visitors to 'take possession of things'; to bridge the gap of ten centuries and make them part of their lives. In that sense the Viking re-enactments, replicas, and events – and indeed also the wrecks – are not really interpreted as 'auratic objects' but as traces of their own genealogy – their own identity.

In one essay, the great creator of fantastic worlds J.R.R. Tolkien asked the question: in what sense is a fairy-story 'true'? In what sense is it 'authentic'? Tolkien's answer to these questions is that it is not really the question of origin (of themes, morals, objects, and so on) that constitutes the authenticity of a fantastic story, but rather the urge to 'get things right' (Tolkien 1997: 131ff). By creating a 'secondary world', the fantastic story offers a complete and coherent universe to step into, engage with, and explore. The desires that are invested in the fantastic stories are not desires for the 'truth'. They are desires for particular skills (crafts or warrior practices) and character traits (good, violent, patient, and so forth). Engagement with these 'secondary worlds' is not unlike the fascination one sees among enthralled sports fans, apparently deeply involved in an ongoing game that seems completely nonsensical (and hence inauthentic) to those who do not know the game's rules. The fascination with (belief in) the game is only understandable in relation to the whole of the 'secondary world' of the game itself. Thus fantasy and realism are intertwined, especially when the imagination emerges from the inspiration of a series of bodily encounters – many but short – with 'real' objects that are no more 'real' than the products of fantasy.

In that sense the ludic aspects of the way visitors interpret and perform heritage at the Viking Ship Museum are crucial to their connectivity with the world of the Vikings. We therefore suggest the notion of *fantastic realism* to capture how the reading of Viking traces intersects with both the workings of fantasy and the world of corporeal, physically encountered reality. The work of fantasy and the objects on display meld and produce a sense of reality that enables visitors to appropriate the past and make it part of their lives and identities. The playful approach to the Vikings produces identifica-tions – such as masculine, even sexist storylines (like the one quoted at the head of this chapter) about Viking warriors, invaders, brave men, and long voyages; but also about women, households with children, agriculture, clothes, and local trade.

The fantastic realism involved in shipbuilding, sailing replica ships, dressing up, playing Viking games or re-enacting warrior practices tells a story about pleasure *and* meaning, fun *and* historical imagination. Our interviews with tourist visitors to the Viking Ship Museum, from Americans searching for ancestors to Danish families with children, expose fantasy productions and performances of Viking skills, characters, and geographies, where pleasure and the formation of meaning and identity go hand-in-hand.

Viking trails

As we have seen, popular imagination and the search for origins are important elements in people's interpretation of heritage places. The fantastic realism of the Viking world facilitates an imaginative repossession of the past. To some extent this reflects a nostalgia for the place of origin; a nostalgia for the 'roots' of ancestors comparable to the 'genealogical tourism' of Irish Americans, for example (Nash 2003). However, the popular genealogies put in play by people visiting the Viking Ship Museum are tied not only to such place-bound roots but also to the routes taken by the travelling Vikings.

While 'the Vikings', their military victories and their territorial claims, became important signifiers for the nation-state projects of the nineteenth century, this nationalization of Viking heritage has been challenged in recent years. In its early national Romantic versions, 'the Vikings' were constructed from an idiosyncratic blend of historically accurate archaeological relics (rune stones, ships), archaeological items from other prehistoric cultures in the region (such as *lur* horns), pagan mythology (the Icelandic sagas and common European fairytales) and sheer fantasy (the horned helmets). During the early twentieth century the Viking iconography progressed from signifying nationhood (as is still evident on coins, passports, and so on) to functioning as symbolism in the labour movement and associated contexts. During the 1940–45 German occupation of Denmark, Viking iconography was used as the central signifier both by the National Socialist movement there and by the Danish Resistance. Similarly, the central icons of Thor's hammer and the *lur* horn were (and still are) used as indications of quality on industrial and agricultural goods (Adriansen 2003 vol. 2). Similar stories can be told about Swedish, Norwegian, Icelandic, and German nationalism. Despite such nationalist appropriations, 'the Vikings' have an appeal that knows no national borders. Of course one is invited to join in 'on the trail of the Vikings' in Denmark, if one arrives as a tourist in Denmark; but along the highway just outside Rouen in France, a road sign can similarly welcome you to join in 'sur la trace des Vikings'.

Traces of Viking culture can be found in an astonishing number of places, and much contemporary fascination with them is related to their mobility. Visitors to the Viking Ship Museum repeatedly expressed their fascination with 'what the Vikings were able to do' not least in terms of their geographical mobility. The traces and marks inscribed by the Vikings on the surface of the Northern Hemisphere guarantee their omnipresent status and enable us to appropriate a Viking past. In that sense Viking culture can be called a rhizomatic culture – a culture of flows, connections, intersections, and mobilities rather than of hierarchy, and these are characteristics that resemble the alleged 'deterritorialization' experienced in the twenty-first century.

Thus everyone – or at least, 'white' Europeans or their descendants, – is able to strike out on the trails into this secondary world, to claim the legacy of the Viking world and make it part of their living history. And in this way the scope of the imagined entrance to the trails of the Viking is globalized.

This imaginative geography of the Vikings is furthermore underpinned and mediated by the multiple objects, icons, and images that refer to an imaginative geography of the Vikings – for example the widely circulated maps depicting Viking sailing routes from Scandinavia to the British Isles, Iceland, Greenland, Newfoundland, France, Spain, North Africa, Palestine, Finland, Russia, the Caspian Sea and the Black Sea, Byzantium and Baghdad. Popular films such as *The 13th Warrior* directed by John McTiernan (1999, based on a novel by Michael Crichton) also feed this fascination with Viking mobility. The film casts Antonio Banderas as the Arab courtier Ibn Fahdlan, who accidentally arrives at the misty home of a band of Vikings somewhere 'in the North' and encounters a world infected by monsters who feed on human flesh and evil sorcerers. Thus it has links with both the fantastic elements of the Viking World and the fascination with their travels. So does the constant flow of popular scientific treatments (in magazines and TV) of 'the truth of the Vinland Map', or 'did the Vikings go to Florida (or did they stop in Boston)?' (Logan 1983). This growing body of Viking objects, facts and fictions, landmarks, iconographies, myths and legends, heroic characters, and so on, is all part of a 'prosthetic culture' (Lury 1998) enabling people to take possession of their past; and thus to construct a sense of genealogy and identity from the world of the Vikings.

In this chapter we have suggested the notion of 'fantastic realism' to capture the desires related to Viking tourism and more generally to heritage tourism; desires that enable people to make connections with objects, activities, posters, movies, and so forth, as traces of a genealogical or imagined cultural heritage. By taking possession of these traces, they associate heritage not only with their own identity and life story, but also with a worldwide network of roads, maps, and landmarks. We have emphasized that heritage interpretations are produced by way of contingent networks among objects, knowledge, fiction, fantasy, and so forth. Heritage sites, their markers, and their iconographic representations represent a network of routes to travel along; maps to use and landmarks to navigate from. As we have learned from the interviews, these landmarks may be found in the genealogical history of one's ancestors, in the cultural tradition, and in cartoons or in popular shows on TV. In that sense the fantastic realism of the world of the Vikings represents a 'spatialization' of history that transforms the multiple constructions of the Viking past into a field; an imaginative geography of heritage. In all its diverse forms such as popular culture, school curricula, theme parks, museums, and re-enactment societies, heritage constitutes maps and journeys. These provide traces to be read and trails to be travelled to the world of the Vikings.

List of interviews

1. Middle-aged couple from Hawaii, USA
2. Middle-aged couple from Denmark
3. Middle-aged/elderly couple from Utah, USA

4. 30–40-year-old research librarian working in Copenhagen, living in Sweden
5. Middle-aged couple from Jutland, Denmark
6. Family with two children, Sweden
7. Family with two children (just back from four years in India), Denmark
8. Three-generation family from California, USA
9. Local grandmother with two visiting grandchildren from Jutland, Denmark
10. Family from Iceland
11. Two sisters (20s) from New York and California (originally from Taiwan), USA
12. Five participants in the enactment 'Warrior Practices' project (20s–30s, British and Irish)
13. Couple (30s) from Iceland
14. Six Interrailers, Italy
15. Shipbuilder, staff
16. Sailor, staff
17. Guide (student), staff
18. Guide (student), staff
19. Tinna Damgård-Sørensen, Director of the Viking Ship Museum.

Notes

1   We would like to emphasize the inspiration we have drawn from the voyages of Bjarke, Gry, and Liv, who have navigated the strange waters of the secondary world; thus in many ways both anticipating and outdoing the endeavours of their fathers.

2   The field work included participant observation and interviews at the Viking Ship Hall and the Museum Island, Roskilde, Denmark. We would like to thank all the interviewees – director, staff and visitors – at the Museum.

Chapter 9

Art exhibitions travel the world

Chia-Ling Lai

Introduction

In the era of global mobility, when visitors travel abroad to museums renowned for their admired artworks, they may be disappointed to find their favourite paintings missing from their galleries. This is because museum artefacts also travel, through art exhibitions linking museums across borders. The protagonist of this chapter is the international travelling exhibition (ITE), which is composed of exhibits temporarily loaned from museums and galleries in a number of countries and usually organized by curators from at least two countries. As a temporarily cross-national exhibition, the ITE can create spectacular events attracting many visitors. Moreover, those cities able to host ITEs or lend major travelling exhibits to such events demonstrate their cosmopolitan credentials and enrol their people as global cultural citizens. ITEs signal a new era for museums and heritages, not only because of their intensified global connectivity, but also their new relationship with society. However, ITEs have received little academic attention, nor substantial academic attempts to elaborate the translation and connection they foster between museums and global heritage, and how this might relate to contemporary tourism mobilities.

This chapter therefore first explores the historical rise and global translation of ITEs and provides an overview of their current worldwide pattern. By looking at museums in Taiwan, I examine the condition of global translation and how ITEs have emerged as an important strategy in the museum field, facilitating interconnectivity between museums cross-nationally. Finally, I outline the multifaceted characteristics of ITEs to shed new light on the politics of global mobility and the transformation of global heritage.

The international travelling exhibition emerges on the global stage

Historically, the first ITEs were launched in the 1960s under a French cultural policy based on a democratic gesture aimed at making the greatest works of art accessible to the largest possible number of French citizens, by promoting a series of high-profile international exhibitions (dedicated to Picasso, Tutankhamen, and so on; see Bazin 1967: 277; Lane 2000: 53). In the 1970s, the Metropolitan Museum in New York started a series of spectacular exhibitions of loaned exhibits from abroad, arousing great attention (Barker 1999: 128). Thomas Hoving, the museum's director 1971 to 1980, was the key figure in promoting ITEs. Interviewed in 1998 by *New York*

Magazine, he recalled that hosting ITEs at the Metropolitan was a response to charges of elitism, and had the aim of making the museum more welcoming, especially to younger visitors.

He also explained that the idea of mounting ITEs was indeed inspired by the European museums' huge art shows. Surprisingly, his interview shows the ITE's two-directional cross-national linkage between museums: lending a collection for an ITE generates the incentive to host ITEs from other museums in turn. Explaining this mechanism of equal partnership, Hoving recalled, at that time,'[European museums] were always borrowing from the Met but never lending to the Met. I said "…from now on, we're going to be partners", so we'd say…No loan until [the European museums] allow us to have some pieces for the X show' (Gross 1998).

These so-called 'blockbuster exhibitions' are characterized by massive media appeals and visitor attractions. They are often analysed in relation to the marketiza-tion effects of museums that operate in an environment without any (or without sufficient) government funding (Wallis 1994; Barker 1999). However, both the French and the American experience show that simply considering the special exhibi-tion as a 'blockbuster' cannot fully explain why museums short of funding would risk mounting such high-profile exhibitions, loaned from abroad, since they actually require huge expenditure.

Replacing this concept with that of the 'International Travelling Exhibition', I suggest that this sort of exhibition can only be understood in the context of the trans-formations that have taken place in the museum field in an era of global mobilities (Bourdieu 1993; Urry 2000), of which the emergence of ITEs was an unexpected consequence. Like a bridge, ITEs connect museums and facilitate links between them across national borders. The ITE's fundamental characteristic of a gift exchange (Benedict 1983) stimulates museums to lend out their own collections in order to have the chance to host similar exhibitions in the future. However, hosting an exhibition involves sustaining certain scales of production and consumption. For instance, only when the Met obtained abundant financial and cultural capacities in the 1970s could ITEs become possible in the United States (Hoving 1993). Moreover, once the domi-nant museums had set the rules of the game, their lesser counterparts tended to follow. European museums garnered more symbolic power than US ones in the 1970s, so American museums imitated and adapted the successful European exhibition model.

Following the same mechanism, in the 1990s, numerous large museums, not only in Europe and America but worldwide, held ITEs. Based on the *Art Newspaper*'s annual survey of the most popular art shows in the world from 1997 to 2001, and related reports on the internet,[1] the general pattern of ITEs in current museums around the world can be seen (Tables 9.1 to 9.4). First, travel has become the most crucial charac-teristic of popular exhibitions. Through their multi-destination travels, these exhibitions link heritages and museums with cross-border visitors. American ITEs tend to travel throughout the United States – for instance, 'Monet and the Mediterranean' was shown in the Kimbell Art Museum in Fort Worth, Texas, and in the Brooklyn Museum of Art in New York (Table 9.1). Numerous ITEs crossed the Atlantic rim and were displayed both on America's East Coast and in London. For example, a constant flow of exhibitions travels between the National Gallery in London and the Met. Apart from the long-term relationship between American and British museums, museums in other areas also cooperated with each other. For

Table 9.1 Top-ten most popular museum exhibitions in the west, 1997

Rank	Visitors per day	Total Visitors (1000s)	Exhibition	Venue
1	6,042	489	Renoir's Portrait	Art Institute, Chicago
2	4,500	434	Picasso and the Portrait	Grand Palais, Paris
3	4,424	531	Picasso: Early Years	National Gallery of Art, Washington
4	4,420	372	Georges de la Tour	Grand Palais, Paris
5	4,318	220	Art in the 20th century	Martin-Gropius-Bau, Berlin
6	4,027	335	Monet and the Mediterranean	Kimbell Art Museum, Fort Worth, Texas
7	3,500	255	Monet and the Mediterranean	Brooklyn Museum of Art, New York
8	3,240	165	Art and Anatomy	Philadelphia Museum of Art
9	3,277	29	Maharaja	Pralormo Castle, Turin
10	3,217	270	Art in Vienna	Van Gogh Museum, Amsterdam

Table 9.2 Top-ten most popular museum exhibitions worldwide, 1998

Rank	Visitors per day	Total Visitors (1000s)	Exhibition	Venue
1	5,339	460	Van Gogh's Van Gogh	National Gallery of Art, Washington
2	5,230	566	Monet in the 20th century	Museum of Fine Arts, Boston
3	5,226	428	The Collection of Edgar Degas	Metropolitan Museum, New York
4	4,461	301	The Art of Motorcycle	Guggenheim Museum, New York
5	4,362	410	Gianni Versace	Metropolitan Museum, New York
6	3,296	300	Alexander Calder: 1898–1976	Museum of Modern Art, San Francisco
7	3,288	306	Delacroix: The Later Work	Philadelphia Museum of Art
8	3,029	300	China: 5000 Years	Guggenheim Museum, New York
9	3,000	302	Rene Magritte	Royal Museums of Belgium, Brussels
10	2,938	253	Recognizing Van Dyck	Philadelphia Museum of Art

Table 9.3 Top-ten most popular museum exhibitions worldwide, 2000

Rank	Visitors per day	Total Visitors (1000s)	Exhibition	Venue
1	6,843	630	Greco: Identity and Transformation	National Gallery, Athens
2	5,876	570	Earthly Art – Heavenly Beauty: The Art of Islam	Hermitage, St Petersburg
3	5,495	500	Sinai, Byzantium, Russia	Hermitage, St Petersburg
4	5,002	355	Seeing Salvation: Image of Christ	National Gallery, London
5	4,290	386	Picasso's World of Children	National Museum of Western Art, Tokyo
6	3,892	280	Dutch Art: Rembrandt and Vermeer	National Museum of Western Art, Tokyo
7	3,879	283	Amazons of the Avant-Garde: Six Russian Artists	Guggenheim Museum, Bilbao
8	3,808	594	The Glory of the Golden Age	Rijksmuseum, Amsterdam
9	3,762	316	Van Gogh: Face to Face	Museum of Fine Arts, Boston
10	3,706	315	Van Gogh: Face to Face	Detroit Institute of Arts

Table 9.4 Top-ten most popular museum exhibitions worldwide, 2001

Rank	Visitors per day	Total Visitors (1000s)	Exhibition	Venue
1	8,033	554	Vermeer and the Delft School	Metropolitan Museum, New York
2	7,178	560	Jacqueline Kennedy: The White House Years	Metropolitan Museum, New York
3	4,924	862	The Medici and Science	Galleria degli Uffizi, Florence
4	4,403	423	Renaissance Italy	National Museum of Western Art, Tokyo
5	4,240	318	Van Gogh: Face to Face	Philadelphia Museum of Art
6	4,151	397	Frank Gehry	Guggenheim Museum, New York
7	4,062	283	Giorgio Armani	Guggenheim Museum, New York
8	3,992	283	Mies in Berlin	Museum of Modern Art, New York
9	3,950	442	Modern Starts: Open Ends	Museum of Modern Art, New York
10	3,868	279	Impression: Painting Quickly in France	Van Gogh Museum, Amsterdam

example, 'El Greco: Identity and Transformation' travelled to museums in Athens, Madrid, and Rome. The Guggenheim Museum's worldwide branches also present a rising significant chain of ITEs, such as 'the Art of Motorcycle' and 'China: 5000 years', which travelled to Guggenheim museums including New York, Bilbao, and Berlin (Table 9.2).

Second, ITEs are usually 'co-produced' by many museums across national boundaries. 2000's 'Earthly Art, Heavenly Beauty – The Art of Islam' was co-organized by the Benaki Museum, Athens; Nieuwekerk Museum, Amsterdam; Islamic art of Nasser D. Khalili, London; the Manuscript House, Sinai; the Metropolitan Museum, New York; and the British Museum, London (Table 9.3). Cross-national cooperation between museums has profoundly transformed the way museums and exhibitions practise. Also, the museums that regularly host ITEs still tend to be the largest museums in their countries (Tables 9.1 to 9.4). Half of the ten most visited exhibitions in 1997–2002 are mounted in museums in the United States. The museum world is still a hierarchical universe in which hosting ITEs remains the privilege of certain dominant museums in specific countries, especially the United States.

Third, although the ITE repertoire remains limited, compared to the avant-garde Biennale in Venice, Madrid, and Taipei, many newly invented topics and novel perspectives have emerged. Impressionists and Post-Impressionists such as Monet and Van Gogh remain dominant, but are viewed from varying perspectives, as suggested by such exhibition titles as 'Painting Quickly In France', 'The Collections of Edgar Degas' and 'Monet and the Mediterranean'. Less well-known painters appeared in the repertoire, such as Georges de la Tour. Ancient civilizations remained a dominant subject of ITEs, not only Egypt but also Islam, China, Byzantium, Sinai, Russia, and Renaissance Italy. Clothing and architectural designers gradually entered museum repertoires through ITEs such as Versace, Gehry, Armani, and Mies van der Rohe (Table 9.4). The convergence between different genres of museum can also be seen, in 'Art and Anatomy' and 'The Medici and Science' (Tables 9.1 and 9.4). Also, new artists emerged with cross-national links and topics relevant to different countries. For instance, 'Mies in America' and 'Mies in Berlin' stressed the architect's cross-continental career in both places. El Greco's life trajectory through Greece, Spain, and Italy made him an extremely attractive ITE topic in 2000.

Fourth, although critics predicted that ITEs have passed their heyday and will soon decline, they are still flourishing, with high visitor numbers. These turn ITEs into global tourism events that pull tourists towards exhibition host cities during a particular timeframe. During the exhibiting period of 'Vermeer and the Delft School' some 8,000 visitors a day went to New York's Met (Table 9.4). Around 6,000 visitors per day went to the National Gallery, Athens, when 'El Greco' was on. Outside Europe, the one-year US tour across three museums of 'Van Gogh: Face to Face', from Amsterdam's Van Gogh Museum, attracted about one million visitors altogether (Tables 9.3 and 9.4). Furthermore, these popular exhibition topics often appear in similar forms in many other exhibitions, thus multiplying the social momentum around the world. For example, the El Greco exhibition had several variations shown in succession around the world. After the shows in Athens, Madrid, and Rome in 2000, exhibitions with various 'El Greco' themes were also shown at New York's Frick Collection and the Kunsthistorisches Museum, Vienna in 2001. Some museums, such as the Met and Guggenheim Museum in New York, still play important roles in initiating new ITE topics. Through their dominant symbolic influences, many other leading museums around the world imitated and adapted similar issues in various forms. Through this dynamic of global translation, these museums coin new global cultural currencies that are accepted by others around the world.

Dos Santos's research also shows that in the 1990s, Brazil also began hosting ITEs based on great artists, loaned from abroad and organized by foreign curators with international standards. They were held in large museums, received great media attention, and many visitors. For instance, the Monet exhibition mounted in Nacional de Belas Artes attracted 435,000 visitors, which outnumbered general visitors to Brazilian museums by thirty times (Dos Santos 2001: 38). Taiwan, discussed below, demonstrated a similar pattern in the 1990s. The ITE has now permeated most corners of the developed world, connecting museums not only in Europe and the United States, but also in Asia and South America, powerfully manifesting the global mobility of museums and heritages, and the mobilities of people and of the objects they instigate.

Translating the ITE into Taiwan: conditions and dynamics

How then did ITEs come to be integrated as cosmopolitan cultural events into Taiwan's museum practices? When the Republic of China in Taiwan lost its seat in the United Nations in 1971, Taiwan was disconnected from official international cultural communications. However, Taiwan's re-emergence on the world economy in the 1980s contributed to its cultural reconnection to the world (Howe 1998). At first, western cultural signs permeated Taiwan at accelerated speed through both imported commodities and global cultural industries including music, film, and advertising, becoming ever more endemic as the economy became steadily more service-oriented. Rapid growth in higher education also provided many consumers with the basic cultural capital to consume western cultural signs (Ministry of the Interior 1994). The deregulation of newspapers in 1989, and the rapid increase in outbound tourism since the late 1980s (Huang *et al.* 1996), further increased the flow of information about foreign museums and Western art through both images and touring experiences (Fang 1993). Finally, as new Taiwanese capitalists emerged in the 1990s who were less deferential to the state, they took up art collecting and the sponsorship of cultural activities as elements of their struggle to establish their own sphere of influence in Taiwanese society. By exhibiting celebrated art collections, they connected themselves to the circle of global capitalists. Competition between these newly emergent capitalists was often played out in the arena of private sponsorship of cultural activities such as ITEs, which they believed could strengthen their image as pioneering, cosmopolitan, and accessible people (DEOA 1995).

These external structural transformations contributed to the circumstances in which ITEs began to come to Taiwan, but only the dynamics of struggle in Taiwan's museum field anchored the survival of this trend, and steered its direction. Two groups of museums dominate the field: one traditional, the other contemporary. After the Kuomintang were ousted from mainland China to Taiwan in 1949, a group of museums were established in Taiwan to conserve Chinese culture. They were built in ancient Chinese style and housed abundant collections of stunning quality. The National Palace Museum is the finest representative, dominating the field in presenting work to foreigners and educating citizens.

The second group of museums, built in the 1980s, focused on presenting contemporary Taiwanese cultural achievements and introducing international trends. Taipei's Fine Art Museum is a typical case. Housing a collection of very limited size, it emphasizes presentation and education more than conservation. However, speaking in the language of modernist architecture and presenting contemporary art exhibitions, these new museums were more accessible than their older counterparts. Since their stated purposes included introducing international developments, when ITEs expanded their influence to Asia these museums became pioneers in establishing links with foreign museums. In the late 1980s, when some low-profile ITEs were launched, their visitor numbers were no less than those of their grand traditional predecessors (ROC Tourism Bureau 1996).

Social change in Taiwan in the late 1980s added new variables to the struggle of these two museum groups. When Martial Law was lifted in 1987, the opposition, pro-independence party formed and won seats in parliament (Clough 1998: 15), which changed the structure of museum funding. The traditional museums were soon charged with elitism and neglect of their educational function, and were threatened with funding cuts. Popular appeal thus became essential for all Taiwanese museums. The contemporary museums' strategy of mounting ITEs was soon adopted by the traditional museums. Some rising young capitalists, proud of their Western art tastes, began to sponsor these exhibitions; their own art foundations actively forged foreign connections for these traditional museums. Gradually, hosting ITEs came to be considered a good strategy in the museum field in the Taiwan of the 1990s.

In travelling to and from Taiwan since the 1990s, ITEs have helped Taiwan become connected to the world museum-scape, yet the global flows of both art exhibitions and the tourists they engender have also been reshaped by their particular localization in Taiwan. Here I consider how Taiwan's ITEs (both outbound and inbound) have enabled the construction of bridges not only to Europe and the United States, but also to China and other parts of the world.

In terms of outbound ITEs from Taiwan, except for the 'Chinese treasures' in the National Palace Museum (NPM), which have gradually become part of one global brand of ancient civilization, few other museums in Taiwan hold equivalent world-class collections suitable for ITEs. Though some travelling exhibitions from the NPM to US museums had occurred before 1970,[2] the severing of diplomatic relations between Taiwan and the international community (as most countries switched their diplomatic recognition from Taiwan to the People's Republic of China during the 1970s and 1980s) prevented ITEs from being implemented again until after 1990. In 1991, Washington's National Gallery organized the exhibition 'Circa 1492: Art in the Age of Exploration', surveying the world's visual cultures around the year 1492. The NPM was invited to contribute, since some of its collection precisely met the exhibition theme. This was the NPM's first loan of exhibits for thirty years, and it initiated a series of overseas journeys for NPM exhibits, such as the exhibition 'Splendours of Imperial China', which went to four leading American museums in 1996, 'Memories of the Empire' to Paris in 1998, and 'Treasure of the Son of Heaven' to Berlin and Bonn in 2003.

Taiwan has also been intensively involved in hosting ITEs since the 1990s. The first few cases were held in the new group of large museums, such as the Taipei Fine Art Museum (TFAM). Their contemporary and international orientation made them suited to ITEs already established in Europe and the United States. However, the connections to museums abroad, according to TFAM's previous director, Dr Huang, were strategic associations accomplished only with great effort. He recalls his way of getting TFAM's 'cultural passport':

> My initial approach in making foreign connections in order to mount ITEs can be summarized as 'from small to large'...I believe it is best to begin by approaching foreign museums in small towns or remote places; relations with museums in larger cities can then be developed from this base...My next principle is to begin such a relationship with 'small volume, good quality'...if we choose to send only a small number of items from a good collection, we can achieve our goal more easily. Then...when the museum abroad has an ITE project, or such an exhibition is already underway in a nearby country, it is a great chance for me to make a connection with that museum and their project. It is surely the case that...our museums' own capacity and reputation in making international exchanges will also increase foreign museums' interest in connecting with us.
>
> (Huang 1997: 103, my translation)

According to Huang's argument, it is no surprise that the first few ITEs held in TFAM featured inexpensive reproductions and minor works of great artists, in exhibitions such as 'Van Gogh's Footpath' and 'Miro: Spirit of the Orient'. Despite this, they still attracted many visitors, and for the first time generated the novel spectacle of queues at Taiwanese museums.

'Monet and Impressionism' travelled from the Musée Marmottan in Paris to the NPM in 1993, showing that ITEs had become accepted as essential by the most influential Taiwanese museums. Its surprisingly large attendance, 310,000, and the astonishing scale of related media reports and cultural analyses it aroused, confirmed its social significance. The NPM's 1997 'Louvre Special' signalled the successful direct connection between large world-class museums and those in Taiwan. For this exhibition, the NPM even launched a new wing for accommodating these 'foreign guests'. It also facilitated further exchange between French and Taiwanese museums.

After 'Monet and Impressionism', 'The Golden Age of Impressionism' from the Musée d'Orsay in Paris, held in the National Museum of History and Kaohsiung Fine Art Museum in 1997, attracted over one million visitors (Chang and Yin 1997), and was followed by the 'Collection from the Musée de l'Orangerie des Tuileries' at the TFAM and KFAM in 1999–2000. The 2002 exhibition 'From Poussin to Cézanne – 300 Years of French Painting', combining exhibits from 27 museums in France, marked a coming of age in cultural communications between France and Taiwan; Western art exhibitions in Taiwan were also no longer limited to the Impressionists.

After 1998, ITEs from mainland China flourished in Taiwan, from 'Sanxingdui', 'The Terra Cotta Warriors and Horses of China's First Emperor', and 'Tunhuang', to 'World of the Heavenly Khan'. These exhibitions were considered not only Chinese treasures but also global brands. For instance, before showing in the NPM in 1999, 'Sanxingdui' had travelled to London, Tokyo, and New York (Fang 1998: 282).

Apart from China's ancient archaeological heritage, many other ancient world heritages, such as those of Egypt, the Maya, Mesopotamia, and India, have emerged in the ITE repertoire in Taiwan. ITEs brought a global heritage of art and culture that had not previously been imagined or witnessed there before. Though still predominantly built on Taiwan's existing links to European and American museums and heritages, which have actively constructed a new cultural world map, museums in Taiwan are beginning to make their own connections outside the dominant West European and American museums. These mobilities of cultural heritage thus have the power to transform global networks and global flows of people.

Characteristics of the ITE

This study demonstrates five characteristics of ITEs in Taiwan. First, they involve cross-national strategic associations between museums and cross-border networking. They embody a new social mobility in terms of horizontal fluidities (Urry 2000: 3). These 'horizontal mobilities' are considered strategies for upwardly 'vertical mobilities' within the museum field. They therefore both exhibit the museum field's globalization and contribute to that globalization. Thus, rather than being random, networking between museums depends on the 'worldview' of a specific museum, based on the quality and extent of its collection (which in turn depends on the historical wealth and power of its host country). The best-endowed museums therefore have a position of dominance in their field that symbolizes their country's past prestige (even if not their current situation). This strongly informs the cross-border cooperation and competition between museums, contributing to ongoing symbolic struggle. As a field, the museum has its specific forms of capital, including quality and quantity of collection; aesthetic and cognitive display capacity; and conservation and research skills. Different museums occupy their relative positions in the field according to the forms and volumes of these kinds of capital that they possess. However, besides some leading 'world museums', the hierarchical relationship between museums in different countries is less rigid than that between different museums in the same country. Thus, it leaves more room for 'unequal' strategic associations that provide resources – particularly for the less powerful partner in the relationship – for ongoing struggles between museums domestically and internationally.

ITEs are not simply the co-product of the two or more of the museums involved. They travel. ITE mechanisms of departure differ from their means of arrival. Like the World Expo, where host cities demonstrate their power to gather exhibits loaned from abroad, guest museums show their capacity by presenting their own exhibitions overseas (Benedict 1983). Also, ITEs can be viewed as gift exchanges between different countries (Benedict 1983). Since museum objects often symbolize the identity of social groups, sending out and receiving exhibits involves subtle protocols of exchange. Invitations, equivalent returns, and the rhetoric of presenting become significant in the process of making connections. However, ITEs need different kinds of museum capital from other common exhibitions. These include a location attracting people and media; fund-raising capacity; the ability of their curators to conceptualize, and of their marketing departments to sensationalize, exhibitions; and the capacity to co-ordinate and associate with other museums and institutions. As the ITE becomes a new arena for museums around the world to struggle in, its proliferation transforms and reorders the structure of the global museum field and of global travelling.

The second characteristic of ITEs is that they are temporary exhibitions, usually lasting only three or four months in any one location. Unlike an imperialist looting of culture to add to permanent collections, the ITE is a temporary loan that functions as a means of sharing collections while supplying destination museums and expanding the display capacity of the source museums. Timing becomes an exploitable resource. Lending often happens when museums are under reconstruction or particular pieces are not on display, thus capitalizing on 'dormant' resources. To arouse attention, ITEs tend to coincide with anniversaries or festival days, thus time also becomes potentially political. The museums' exhibition timetable creates the frame of comparison and competition between different sorts of exhibitions, so ITEs are often held in different museums at the same time.

The ephemerality of ITEs renders the eternal tempo of museums and heritages compatible with an age dominated by instantaneous time (Urry 2000), where the speed of media news around the world is overwhelming. Potential visitors, both domestic and international, must be concentrated into the short period of the exhibitions, so ITEs compete to attract media attention on the 'global stage' (just as 'global cities' compete to attract tourists; see chapter 12). In a society gradually desynchronized (Urry 2000: 128), the temporary ITE becomes a rare forum for initiating collective organized activities and giving its society a shared 'place to play'. Sometimes, various related cultural activities surrounding the ITE's theme, not only in the forms of visual art, but also in the performing arts and in general leisure entertainments, are all synchronized in the exhibition period and render them multi-bill festivals. They not only offer social topics for local conversations; they also create 'a global form of play' that attempts to create an atmosphere in which individuals and institutions feel they cannot afford *not* to participate. ITEs coin a new 'cultural currency', globally recognizable and convertible, which any cosmopolitan cultural citizen cannot afford not to possess. However, the mechanism of social distinction does not disappear. Instead, visitors no longer separate themselves from others merely by attending or not, but also by competing with each other through the question of when to attend. For instance, to avoid crowds, some enthusiastic visitors even exceptionally take days off work to attend the exhibition; or compete to be earlier visitors in order to become the authority on it within their social group. The ITE as a powerful event restructures the schedule of everyday life and time itself becomes a new dimension of social distinction (see chapter 16 on similar processes of aesthetic distinction within places to play in Dubai).

The third characteristic of ITEs means that museum objects are on the move. In the age of mobility, the movement of objects is by no means novel. However, the museum collection, highly invested with symbolic meanings, can be difficult to move. Museum objects are classified by Lury as *traveller-objects* that retain their meanings and authenticated relation to an original dwelling when they cross contexts. Typically they stay still, under legal restrictions on movement across national boundaries, although their images frequently move (Lury 1997: 78). However, through ITEs, these normally static traveller-objects are temporarily allowed to move across national borders. I intend to twist Lury's argument further: while some museum exhibits cannot move, others can be moved. Their mobility depends not merely upon professional classification and practicalities (such as the size and weight of objects, their fragility and preciousness, or their meaning to particular indexes of collection). It also depends upon the perception of and attitudes towards these precious exhibits created by insurance coverage and how well the museum appears to care for them.

Since a hierarchy among collections exists in terms of their 'mobility', the ability to initiate the departure of the most 'immovable' precious artefacts from loaning museum becomes a significant manifestation of the power of host museums. The hierarchical relationship between museums on the international scale sets conditions on museums' abilities to appropriate objects from other museums abroad. To some extent, hosting ITEs resembles the 'temporary collecting ability' among museums as collectors. Friendly delivery of a museum collection abroad is often regarded as a kind of democratization, since it circumvents physical and social distances that might hinder cross-national museum visiting. Indeed, ITEs remove some physical, if not cultural, inaccessibility from museum visiting (Bourdieu and Dabel 1991), insofar as they may replace travel to the loaned museum abroad. But more often they serve as promotional introductions arousing people's desire to travel. Some visitors who can travel afar can even enjoy comparing several variations of the same ITE. The multiple mobilities of movable global heritages *and* mobile spectators thus create the complex dynamics and context for a new politics of global mobility.

Fourth, the ITE is a convergence of exhibits and a point of contact between museum practices. As researchers on World Expos (Benjamin 1978; Rydell 1984; Simmel 1997) point out, the attractive spectacle of exhibitions is created by the convergence of objects. Also, the ITE is more a co-arrangement resulting from contact between different museum practices (Clifford 1997), based on various beliefs about museums in different countries. Neither random nor completely predictable, the contact between museums and the convergence of exhibits embodied in the imaginative construct of the ITE is intertwined with the various ranks of historically based prestige (alluded to above) invested in the hierarchical relationship between museums. Within that convergence, the exhibition provides comparison, competition, or hybridization between different cultural or art forms. Appealing to expected audiences from various cultural backgrounds, exhibitions must take into consideration cultural perceptions and regimes of value across and between different contexts.

Furthermore, the current 'thematization' of exhibitions emphasizes new perspectives and novel aesthetic styles for organizing and presenting exhibits (Heinich and Pollak 1996), making the ITE a 'generic space' (Lash 2002; and see chapters 5 and 10 on generic tourism spaces). Themes, like brands, create a bounded meaningful arena for a group of selected exhibits – becoming 'logos' that are recognizable across borders and enable the mobility of museum collections over different cultural contexts. Moreover, the theme can become part of a cultural fashion, surviving with variations that are appropriated by different museums. However, 'thematization' requires museum curators who can command conceptualization, visual presentation, and sensuous communication simultaneously, and museum practices and facilities to support it. With the rise of ITEs, the capacity to thematize could come to overshadow the ownership of collections and become a new logic for global heritage struggles.

The ITE's last but not least feature is its packaged mode. Though the ITE is a spectacular generic space providing attractive cultural imaginations, it is also a packaged travelling exhibition arranged by certain sets of regulating principles, much as the package tour is 'McDonaldized' (Ritzer and Liska 1997). First, as a rare chance to gather numerous masterpieces, with spectacular and delicate designs, together with various simultaneous visual- and performing-art activities as adjuncts, ITEs offer no standard menu, but rather a tantalizing 'once-in-a-lifetime' multi-bill feast for consumers to attend. Further, it also serves as an 'efficient' way to consume at once a well-designed summary of an artist's whole life journey, or the plentiful cream of foreign museums' collections. Also, though the huge number of visitors becomes an unpredictable factor for such exhibitions, nowadays, techniques of delicate 'control' of visitor flows have been developed to divide the visitors into several 'flows', limiting viewing during certain times and requiring advance booking.

This new style of exhibition, differing from usual museum exhibitions, requires new cooperation throughout many sectors, including insurance, transportation, architectural design, and mass media. The cross-national flows of objects, meanings, and practices through the ITE involve a strong network of expenditure, organizational and interpretation effort, and time. The huge costs of insurance and display renders the ITE an extraordinary project needing extra sponsorship. The spectacular design of the 'themed exhibition' requires further support from the exhibition's architectural designers and project organizers, who might come from outside the museum field. ITEs usually take much more time to produce (three to five years) than general museum exhibitions. The cross-national exhibition implies a move into a different regime of value, requiring large audiences to realize the necessary income within the short period the exhibition is in any one location. Therefore, wide media coverage becomes crucial for the ITE.

Such exhibitions also require coordination between many other cultural performances, such as dance, music, theatre, film, and lectures; as well as leisure activities, such as restaurant and tour packages, and promotional activities, such as prize draws and souvenirs. Also, when attracted to witness the exhibition, visitors render themselves a part of it. With large numbers of visitors, the exhibition takes on something of the character of a festival. New kinds of 'de-differentiation' between the genre boundaries of art, and between high and popular culture, have emerged within the ITE.

A central new task in ITEs is risk prevention. The movement and display of artefacts is often mentioned as a risk; the required large attendance constitutes unpredictability. Apart from safety concerns of transportation and display, the spectacular exhibition and its expected social response must also be well planned in advance and carefully guaranteed during the exhibition period. Besides the museum itself, many other institutions, such as insurance companies, might be involved. However, the involvement of insurance would transform our understanding of culture and its mode of values, since the aesthetic value placed on an artwork tends to reflect its monetary value as reflected in an insurance valuation. Therefore, these multiple coordinations between museums and other institutions, based on supporting the new task, appropriate the expected symbolic capital created by the ITE, and transform museums into new places to play.

Conclusion

Co-produced by museums cross-nationally, gathering and presenting museum objects across borders and connecting museums around the world, the ITE constitutes a world stage that museums cannot afford to miss hosting and presenting. Delivering the cream of the world's museum collections in an approachable way, creating a sensational, efficiently packaged feast of world culture, shared by people in many corners of the globe, the ITE has made itself into a global event.

This chapter traced the historical emergence and current pattern of this global event on the world stage and analysed the condition of its indigenization by examining museums in Taiwan. I have shown that ITEs provide a worldview regulated by previous forms of symbolic domination, but also contributing to emergent 'worldscapes' of museums and heritages. Finally, I suggest that in the era of global mobility, global heritages no longer perform just in a fixed site. Instead, they perform globally (and generate new tourism mobilities) through art exhibitions that constantly travel to connect, strategically, their foreign counterparts. Tackling the risks inherent in coordinating complex packages of sponsorship, insurance, and transportation, ITEs mobilize spectatorial travel both nationally and internationally through their dynamic tempo and 'must-see' status. These multiple mobilities of objects of global heritage and visitors at large constitute the new politics of global heritage.

Notes

1   The ephemeral exhibition is given long-term material form in the lavishly illustrated catalogues that are published to accompany them, and to a certain extent on the internet. Tracing each top exhibition on the internet, I examine its itinerary and the content of the exhibition.
2   The ITEs from Taiwan in the 1960s and the 1990s are very different: the 1960s exhibitions were carried by warship, influenced heavily by its political duty, and took one and a half months to travel. The visitor numbers were about 100,000 in each venue (Lee 1972). The 1990s exhibition carried by airlines took just one day to arrive. The visitor numbers in one museum, the Metropolitan Museum of Art, are about 426,000.

Chapter 10

Reconstituting the Taj Mahal: tourist flows and glocalization

Tim Edensor

Introduction

The Taj Mahal – the mausoleum built during the 1630s and 1640s by the Mogul emperor Shah Jahan for his favourite wife, Mumtaz Mahal – is not only a clichéd symbol of India in the global 'image economy', but has become a truly global icon used to conjure up notions of planetary unity and diversity, along with other super-symbols such as the Statue of Liberty and the Egyptian pyramids. Thus the image of the Taj flows across global mediascapes and is ubiquitous in its familiarity. Such image and knowledge flows emerged during the colonial era through the widespread dissemination of British travel accounts, and such flows have thickened ever since. In the early 1990s the Taj served as the backdrop to the lovelorn Princess Diana following her separation from Prince Charles, and as one of the symbols destroyed by aliens in the movie *Mars Attacks*. This continuous re-marking of the Taj serves further to underline its importance as a must-see global tourist attraction and consolidate its centrality to tourism in India, and to the city in which it is located, Agra, in Uttar Pradesh.

In my original study, *Tourists at the Taj* (1998), I argued that the monument was spatially woven into a plethora of routes, dependent upon specific themes and variously scaled imagined geographies, and that these itineraries informed the specifics of how a visit to the Taj was practised and how the site was interpreted. Thus Muslim pilgrims include the Taj as part of a voyage incorporating the tombs of Sufi saints and other sacred Islamic destinations, and hence reinforce its Islamic sacrality, a factor that informs their reverent understanding of the site and their practices of walking, gazing, and remembering. On the other hand, for young Western backpackers, a visit to the Taj is generally something of a detour from the backpacker circuit, which typically features sites of hedonism (Goa, Manali), Hindu sacrality (Rishkesh, Pune), and more mundane places identified as part of the 'real' India.

The Taj has been the subject of a process that has seen the local become ever more tightly drawn into the global while the global has become inextricably concerned with the local particularities of Agra (which this chapter will discuss in some detail). Urry refers to this process as 'glocalization' – that is, 'globalization-deepens-localization-deepens-globalization' (2003a: 85), producing inextricable interdependences. Locals get drawn into multiple and disparate processes of the global through seeking flows of information, culture, ideas, money and lucrative flows of people and things, and global flows are drawn towards localities on the basis of their specialities and apparent uniqueness.

The Taj Mahal has been irrevocably glocalized by its weaving into tourist flows, and yet this is only part of the embeddedness of the site within the global. In what follows, I will depict a range of recent changes and controversies that have centred upon the Taj Mahal, and which have had an impact upon India, the city of Agra and its inhabitants, on the tourist industry, and on tourists themselves. These issues are inextricably linked with the intensifying glocalization of this specific attraction and its locale, and they testify to the continual remaking of the Taj, Agra, and tourism. I provide an account of several recent interconnected developments that centre upon discourses and practices oriented around conservation, scientific expertise, economics, the global threat of terrorism, and the development of 'international' tourist standards. I then highlight the ways in which the Taj continually emerges out of a glocal dynamic that generates global pressures and local resistances and accommodations. More specifically, I will look at the flows of global discourses, forms of expertise, economic strategies and adaptations, and the production of changing mobilities, spatialities, and temporalities that centre upon and radiate from the Taj.

Pricing the Taj

In 2000, admission to the Taj Mahal cost 20 rupees (about 30p Sterling) for foreign visitors and 2 rupees for Indians. It was suddenly announced in September 2000 that the entrance fee was to be raised for domestic tourists to 20 rupees and for foreign tourists to 960 rupees (the latter subsequently reduced to 750 rupees by the Supreme Court). These massive increases led to considerable dismay from tourists and locals alike. The decision to raise the fee was taken by the Archaeological Survey of India, who are responsible for the maintenance of the site, and the Agra Development Authority, the rise being justified by the former (who take one-third of the amount) on the grounds that it would finance vital conservation work at the site and at other local historic sites, and by the local authority (who receive two-thirds of entrance revenue) who regarded the extra income as vital to planned social expenditure and the upgrading of local infrastructure, including tourist amenities. It was argued that wealthy foreign tourists could certainly afford to pay for entry and that a potential source of revenue had been previously ignored. Many cynical locals doubt that the revenue will be spent in these ways and charge that the money will be diverted through bureaucratic and political corruption.

Immediately, the huge and unexpected increase had an impact upon visiting Western backpackers who claimed that the charge was excessive and, as budget travellers, well beyond their finances. A group of Japanese tourists staged a sit-down protest, others wrote to local newspapers about what they regarded as discriminatory and 'racist' rates for foreigners in contrast to the lower rate for Indians, and three backpackers were arrested for scaling the outer walls to gain access. At first, it was estimated that 50 to 100 backpackers a day were turning away from the Taj after being unable to afford the entry fee (Harding 2000). As mentioned, the Taj is not generally a primary destination for backpackers, but the compulsion to visit such a renowned spot has been strong. Consequently, Taj Ganj, the area adjacent to the mausoleum complex, has developed facilities for these budget tourists, cheap hotels, restaurants, and numerous shops where they can buy craft and other products. The 'whistle-stop tour' schedule being anathema to backpackers, they had typically stayed in Agra for several days, much to the satisfaction of local traders. Since the increase in admission to the Taj and its disproportionate impact upon this particular group of tourists, however, the reduction in the numbers of backpackers visiting Agra has been severe. Backpackers' itineraries are partly constituted intersubjectively, by the circulation of chat between them in hotels, hotspots, and increasingly on the net, and the huge price hike has become something of a cause célèbre. Even with the subsequent reduction in entry fee to 750 rupees, then, the Taj is no longer a significant element in backpackers' itineraries. Bowcott relates this to wider changes, arguing that the 'heyday of the backpacker is over: nobody publishes guides celebrating the ingenuity of "touring Europe on 5 dollars a day" anymore, while haggling with a rickshaw driver is definitely out' (2001:1). Thus the increase of the Taj entry fee is part of a wider global process mirrored in price rises at tourist attractions across the world where extra revenue is sought to ameliorate environmental and social damage caused through tourism mobilities and to pay for 'improved' tourist infrastructure. In turn, these developments cause the reconstitution of backpacker itineraries as budget tourists seek alternative, cheaper sites around the world.

India has never managed to attract the numbers of tourists that some other Asian countries such as Thailand have. Nonetheless, tourism to India, like many other more popular destinations across the world, was significantly affected by the terrorist attacks in the United States on 11 September 2001, with a subsequent large reduction in foreign tourists. While 250,000 visits to the Taj were recorded in 2001, the following year saw the figure drop to 171,000, although numbers in 2003 almost doubled the 2002 figure. Although annual domestic tourist entries to the Taj are typically six to eight times greater than foreign entries (there were almost 2 million in 2001), these too declined in the first four months of 2002 (figures courtesy of the Archaeological Survey of India) as tension grew between India and Pakistan in their long-running dispute over Kashmir, now heightened by the fact that both countries have nuclear weapons. As large armed forces from either side massed along the frontier, people stayed at home and tourist activities suffered as a result throughout India – even in places such as Agra that are not close to the border. The reduction of tourism in the aftermath of 11 September 2001 and the huge increase in admission charges at the Taj have had severe consequences in Taj Ganj, with the closure of several backpacker hotels and the demise of craft shops. Marble inlay workshops, capitalizing on tourists' appreciation of the *piétra dura* inlay work at the Taj, had previously been the foremost local craft; this has now been reduced to a small trickle and businesses have had to shut down or explore alternative outlets in which to peddle their expertise and products.

In *Tourists at the Taj*, written before the price rise, I wrote about Taj Ganj as being a place in which the local hoteliers, traders, rickshaw operators, shopkeepers, and craftsmen were already indignant about the decline in the numbers of tourists who flowed through their area – the passing trade that was vital to their survival. The capturing and directing of tourist flows by the large operators, as opposed to the more casual itineraries of yesteryear where tourists were apt to wander and organize their own schedules, has meant that the large tour operators have marginalized these smaller enterprises in their attempts to maximize profits. The large tour operators have attempted to confine the movements of tourists through increasingly packaged tours where each day is minutely planned so that there are few opportunities for improvisation. Typically, a visit to the Taj will be surrounded by visits to air-conditioned restaurants, and to a range of up-market craft emporia and other shopping outlets. Scare stories from guides are related to inhibit tourists from voyaging beyond the confines of these interconnected environmental bubbles, thus ensuring that they are more likely to spend their money within approved tourist enclaves (see Edensor 1998, 2001a, 2004b). This controlled production of the tourist 'product' undoubtedly provides a sense of familiarity and security for tourists frightened by the practical difficulties and episte-mological threat posed by encounters with un-contextualized, unmanaged difference. However, while certainly lucrative in the short term for these interconnected busi-nesses, such a system often leads to frustration among the tourists themselves, who often have come to India precisely to witness more 'otherness' than these tours provide.

In Agra, locals complain that too many tours are controlled by large Western tourist corporations and that there is little concern with ensuring that tourists stay *in* the city. This is compounded by what locals refer to as the 'Delhi Mafia', who organize trips to the Taj from the capital, either as day trips or as one-night stopovers in Agra, and thus also limit the potential for tourist expenditure. The construction of the Japanese-managed 'Taj Express Highway' between the new airport at Noida, near Delhi, and Agra to reduce drastically the travelling time between the cities, might further minimize the numbers of tourists staying in Agra overnight. There are plans to erect motels at the side of the road, and the road itself will be bordered and patrolled to prohibit smaller (and slower) modes of transport.

On the other hand, this more comprehensive infrastructure may draw the Taj into a wider regional space of mobility, providing the critical mass that will permit wide scope for varied forms of travel. Yet even if this proves the case, locals say, the incen-tive for package tourists to stay for more than one night in Agra has been largely eliminated by the rise in the admission charge, for whereas tourists were previously happy to pay the small admission charge to visit the monument more than once on successive days, they are now deterred from doing so because of the expense. In the same vein, locals complain that the opening hours of the Taj limit the potential for increasing visitor numbers and the length of their stay in Agra. The site used to be open in the early morning and through the evening so that visitors could gaze upon a moonlit or dawn-lit Taj, as recommended in nearly all guidebooks since the colonial era, and there have been recent suggestions that subtle lighting could be used to simu-late the moon's glow on dark nights to avoid tourist disappointment. However, opening hours are currently dictated by the hours of sunrise and sunset because of increased security concerns.

What is clear is that to try to reverse these trends, there is a need for Agra to diver-sify and develop its tourist product so that rather than constituting the sole attraction, the Taj is the centre of a complex of attractions both near and far. It is often claimed that besides the Taj and Agra Fort, there are historic sites in Agra that few tourists visit, specifically Chini-ka-rauza (the mausoleum of the poet and chief minister of Shah Jahan, Allama Afzel Khal Mullah Shukrullah), the elaborate tomb of It-ma-ud-daulah (chief minister of Jehangir) and Ram Bagh (supposedly the earliest Mogul garden), while nearby are the bird reserve at Bharatpur and the remarkable, abandoned Mogul city of Fatehpur Sikri. The lack of exploitation of these sites and the consolida-tion of vested interests together with a concentration on encouraging consumption-dominated tours – with their attendant rewards to operators in the form of commis-sions from emporia and restaurants – means that Agra is able to capture flows of tourists only over a very limited period of time. And now that the relatively lengthy stays of backpackers are over, Agra is a place through which tourists flow rapidly, although it possesses one of the world's best-known attractions!

Although the economic impact of the rise of the admission fee upon the mobilities and numbers of foreign tourists has been most evident, the rise in the entrance fee for domestic tourists has also discouraged certain groups of Indians from visiting. In short, domestic visitors are now overwhelmingly middle class. Previously, large parties of villagers and poorer groups travelling by bus usually stopped at the Taj en route to sites of pilgrimage such as Mathura and Ajmer, but they are now unable to afford the 20 rupees to gain entry. If they want to visit the site, such parties must now time their visit for the Muslim festival of Urs in Winter, or else arrive during World Heritage Week in April or World Heritage Day in November, when entry is free. Moreover, locals who were able to go into the Taj and used it as a leisure facility are now less likely to do so, and the monument has thereby been somewhat abstracted from their sense of locality.

More controversial, for both local and wider contexts, was the decision that entry would be charged on Fridays, which had traditionally been free so that local Muslims could enter to carry out their religious right to pray at the mosque adjacent to the tomb. This effective debarring of local Muslims led to protests, as a result of which it was decided in 2002 that the monument would be closed to all on Friday, with the exception of those local Muslims prepared to be subjected to security checks upon entry. The tightened security in this case reflects distrust of Pakistani intentions as well as the more ubiquitous perceived global threat of al-Qaida, and dissatisfaction with this solution is also fuelled by complaints that Muslims are being privileged in gaining free entry on a day when entry is entirely closed to others. This in turn feeds into certain Hindu fundamentalist conceptions of the Taj which mark it as one of many sites in which an Indian temple was demolished or transformed to make way for an Islamic structure. Despite the dubiousness of assertions that the Taj was originally built as a Rajput temple (see Oak 1994), it is clear that the global ideoscapes that carry crude depictions of Islamic fundamentalisms feed into a defensive Hindu fantasy in their turn.

Although the transformed flows of tourists through Agra have had an impact on the local tourist industry, particularly on those traders who formerly relied on large numbers of backpackers for their business, it is important to acknowledge that being positioned in a globally dynamic context in which flows of people and money might suddenly dry up requires a reflexive disposition towards the likelihood of change and a flexibility to respond to transformed circumstances. These small traders have had to improvise to both extend their service to tourists and seek other ways of engendering tourist performances.

First of all, new tourist markets have been sought, and there has indeed been something of an influx of South Korean tourists, as increasing wealth has encouraged them to participate in global tourism. This new presence is indicated in Taj Ganj by budget hotels suddenly proclaiming their speciality in Korean cuisine. Nevertheless, given the unfavourable rate of exchange, it is also difficult for South Koreans to afford a trip to the Taj. Since the increase in entry fee, solutions to this problem have been sought through the provisions of other ways of gazing upon the Taj. One strategy has been the development of boat trips across the river to the Mahtab Bagh so that tourists may witness the reflection of the mausoleum in the water, and in other cases, residents close to the site have offered their houses as places from which the Taj may be viewed. In this sense, the tourist pathways around the Taj have been multiplied by these alternative 'gazing' services. Local craftsmen producing marble inlay work have tried to seek alternative markets by using numerous local cyber-cafés to set up websites at very little cost to advertise their products abroad. A perhaps more important factor, however, has been the growth in consumer culture and a status-oriented lifestyle amongst India's growing middle class, perhaps partly stimulated by a growing awareness of and receptiveness to Western ideas of success and happiness. Thus, the demand for marble inlay work, along with other local crafts, has boomed as this middle class complement their quest for Western commodities with the cachet of revalorized traditional artefacts and styles.

Enforcing conservation

> The inexorable growth of foreign tourism, and the importance of heritage, culture and art to that industry, is the most powerful expression of the existence of a common global heritage as the property of all peoples.
>
> (Graham *et al.* 2000: 238)

India is a signatory of UNESCO's World Heritage Convention, established to identify and protect sites of outstanding cultural and natural value throughout the world, and the Taj Mahal is one of the most prominent of these sites. The application to add a site to the World Heritage Sites list comes from the country in which it is located, and it is then incumbent upon that country to provide plans concerning the site's protection, and to pledge to maintain this preservation (see Smith 2003: 99–116 for more details). Accordingly, UNESCO monitors these sites, and intervenes if they feel there are threats to them, the ultimate sanction being removal of the site from the list. But not only is this international body involved in the upkeep of such sites in India, but the powerful Indian Supreme Court is also able to enforce constitutional law (including that enshrined in the World Heritage Convention) and thus also possesses a regulatory function. The Court has most famously intervened to order the closure of polluting factories in Agra which were adjudged to harm the marble surface of the Taj and thus risked broaching India's signature of the Convention.

Before I discuss a later intervention of the Court, I will argue that the glocalization of ecological and environmental agendas, discourses, and practices intrinsic to the mobilization of this World Heritage agenda (and of the Supreme Court as its national agent) are brought to the fore by several other developments around the Taj. Moreover, I will attempt to show that these universalizing glocal debates and policies are supported by certain local and national parties but are resisted and perceived as a threat to the interests of others.

After years of relative neglect, and the subsequent yellowing of the Taj's exterior surface due to the polluting effects of local factories and traffic, the conservation agenda, supported by UNESCO and enacted by Indian conservationists and other actors who regard the Taj as of vital symbolic import to India's tourist trade and overall reputation, is in the ascendant. This preoccupation with ecological and preservation imperatives has led to several recent developments at and around the Taj. The development of plans for a 'Taj Trapezium' – an extensive area of greenery comprising woodland and parkland along both sides of the Jamuna River, intended to insulate the Taj from pollution – is now well advanced, and an area of many square kilometres, within which no development is permitted, is already established. On the eastern side of the monument, a protected forest covering hundreds of hectares is maintained and reforested. Close to the Taj, there is a new 'Taj Walk Park' where visitors may experience spectacular views of the monument as it rises above an uninterrupted vista of tree tops. For foreign tourists, entry to the park requires that they be accompanied at all times by a 'guide' who is responsible for 'security' and ensures that they do not venture into the 'jungle' or come to other harm.

On the west of the monument, the notion of a 'Taj Corridor' aims to link Agra's other major monuments to the Taj within an ecologically designed and protected strip of land connected by footpaths, a scheme that has been partially supported by the advice of gardening specialists from the University of Illinois at Urbana-Champaign (*www.landarch.uiuc.edu/*). By linking the Taj, Red Fort, It-ma-ud-daulah, Ram Bagh, and Chini-ka-rauza, it is envisaged that the tourist allure of Agra will be extended.

In its initial stages, this project has entailed the closure of the bus park used by Indian tourists. This vibrant, noisy place, with numerous food stalls and peddlers, often cloaked in a thick pall of smoke resulting from the noxious emissions of India's notoriously sulphuric buses, and formerly situated about 230 metres from the Taj's western gate, has been resited so that it now lies a further 275 metres away, and the original site has been transformed into a verdant garden. Tourists are now obliged to walk or else take only cycle rickshaws and electric and battery-run forms of transport. Likewise, the buses of Western package tourists are similarly obliged to park some distance from the Taj's eastern gate and take these cleaner vehicles. This is part of a policy that restricts the presence of oil-fuelled vehicles, which may not come within 500 metres of the Taj complex. As well as these green developments either side of the Taj, the Mahtab Bagh (Moon garden) across the Jamuna River from the Taj, is the site of an ambitious project, also informed by advice from the landscapers of the University of Illinois, to recreate a 40-hectare Mogul garden, which will be part of a bigger Taj National Park. The recreation of these original gardens will also supposedly broaden the tourist appeal of Agra, for they may be reached by boat or bus, and the Taj can be viewed, reflected in the Jamuna.

As I wrote in *Tourists at the Taj*, schemes to recreate the Mogul parks and gardens that originally surrounded the Taj also included plans to redesign the gardens inside the memorial compound. These gardens were originally designed to veil the monument with thick vegetation, and contain numerous fruit trees. Early paintings of the Taj show that the tomb was indeed obscured by tall, leafy trees, but the British colonial regime, chiefly in the person of Lord Curzon, 'tamed' the garden so that it conformed more closely to European ideas of perspective and beauty, as found, for instance, in André Le Nôtre's grand vistas at Vaux-le-Vicomte and Versailles. There is currently a vigorous debate within the Archaeological Society of India between proponents of restoring the original garden and those who wish to retain the uncluttered view of the monument. The former advocate the restoration scheme on the grounds of authenticity, whereas those defending the status quo point to the global familiarity of the contemporary aesthetic arrangements. They also suggest that the Mahtab Bagh will provide an example of Mogul horticultural authenticity, and question why the Mogul historical period should be privileged over the British raj. More practically, they suggest that thick foliage would provide a haven for those intent on terrorist activity and furthermore, might conceal the plight of tourists stricken with all kinds of misfortune. Again, this argument underlines the security concerns which have surrounded the Taj in recent years. At the height of the stand-off between Pakistan and India, there were even plans to cover the Taj with a huge tarpaulin to disguise it from potential air attack. In this climate of perceived greater danger, security for the monument has been handed over to the Central Industrial Security Force (CISF), who have intensified the guarding of the Taj and its surrounds.

As far as traffic pollution is concerned, it is intended that all vehicles registered in Agra will be converted to cleaner compressed natural gas (CNG) fuel within three years, emulating Delhi's achievement in combating that city's appalling air pollution problems. In addition to this has been the reinvention and reintroduction of the cycle rickshaw. Long despised as a form of transport exploitative and degrading to its driver, and thus the object of liberal concern for western tourists – and yet always environmentally sound – the cycle rickshaw has been redesigned so it is both more comfortable for passengers and easier to operate, with the addition of gears, greater stability, and the design of easier pedalling and steering. The Taj Mahal Cycle Taxi Improvement Project was initiated by a New York-based organization, the Institute for Transportation and Development Policy, which champions non-motorized, environmentally sound transportation. This has also afforded local industrial opportunities for the manufacture of these vehicles (see Farrell 2002: 8–9).

In carrying out the environmentalist agenda, India's Supreme Court ordered the closure of many polluting factories that were emitting high levels of sulphur dioxide, causing the corrosion of the marble surface of the Taj, unless they were able to provide green methods of energy conservation and emission. The local tourist industry, together with local and national parties interested in conservation, often instigated and supported these actions as a way of supporting their own interests, while they were naturally opposed by local industrialists and workers. The first wave of closures, in 1994, led to vociferous demonstrations from some of the many thousands whose livelihood would be lost, yet a second wave of shut-downs in 2000 did not lead to the same outcry, as many were now resigned to the inevitable clampdown. In response, many factories have relocated within the region but the provision of clean gas from the nearby Mathura refinery has meant that some were able to afford the conversion process that turned them from coal-burning to gas-fuelled enterprises. Yet overall there has been a radical deindustrialization of Agra through the closure of these heavy industries, chiefly foundries producing diesel engines and small engineering parts, and this has forced the economy to adapt to these new glocal realities. Accordingly, there are fast-growing sectors that manufacture gold and costume jewellery, lighter engineering works such as those engaged in the aforementioned production of updated cycle rickshaws, a nascent fashion industry, and greater diversification all round, which has been made possible by greater use of information technology, and has led to an increase in the proportion of women in the workforce.

Despite the reinforcement of the Taj as a World Heritage Site through these numerous developments, a recent controversy has arisen that has potentially threatened this status. As mentioned, the Taj Corridor concept conjures up a vision of an ecological strip linking Agra's historic sites, and yet this notion was recently appropriated by the Uttar Pradesh Government in their plan to erect a commercially oriented corridor in the form of a $40 million complex of shopping malls, restaurants, and amusement parks – to be developed by entrepreneurs (whose identity, at the time of writing, is still not publicly known) – on 29 hectares of land along about 1.6 kilometres of the bank of the Jamuna River, beginning at a point only 300 metres from the Taj site. Funded by the State government through an untendered award to the unidentified company, a large portion of the banks and bed of the Jamuna River was filled with concrete and rubble and the river course altered.

Although shrouded in confusion about which state officials and elected leaders did or did not authorize the project, and whether indeed, there was any organized project beyond an attempt to channel the Jamuna to create a more pleasingly full river for tourists, construction was brought to an abrupt halt by the intervention of the federal Tourism and Culture Minister, Shri Jagmohan, in June 2003, who declared that no permission had been granted for the construction, despite the earlier objections of several statutory bodies. A few days earlier, the Archaeological Survey of India had complained about the project, and crucially, had suggested that UNESCO officials might reconsider the status of the Taj as a World Heritage Site if the project went ahead. A first step would be placing the site on the list of 'endangered' sites, the first step towards withdrawal. Objections primarily focused upon the environmental impact of the project. The diverting of the river and its flood plain, it was alleged, could divert the flow of the Jamuna during monsoon rains and damage the base of the Taj through displaced flooding. The dust generated through construction was also believed to be affecting the surface of the building, and the ambience and 'atmosphere' required at a site of such importance would be negatively affected by the development. Vitally, it was held that India's international reputation would be at stake if it was seen to allow such a globally significant site to be threatened, a sentiment articulated by an investment consultant who claimed that 'foreign investors would have questioned themselves about (the) safety of their investment in a country where even (the) Taj Mahal isn't safe'. The Confederation of Indian Industry also opposed the plans for the same reasons (Singh 2003).

Reconstituting the Taj and tourism

I have identified a number of recent processes and developments through which the Taj Mahal, as an internationally renowned tourist site, has been the centre of various global flows. Since British colonialism in India, the Taj has been embedded within global networks, and this has intensified as its local issues have become ever more tightly and mutually intertwined with national and global concerns – the process of 'glocalization' at work. In ever-changing contexts of international regulation, administration, and flows of global tourism and ideologies, places like the Taj (symbolically loaded with different forms of global, national, and local significance) are likely to be continually subject to change by virtue of the extensive processes that bear upon them. At the Taj, mobilities of capital, governance, expertise, knowledge, ideals, ideas, and people produce temporary spatial configurations that in turn have an impact upon the ways that tourism is performed and experienced by different groups. In order to bring these processes to the fore, I will identify three interconnected glocal processes that continually reconstitute the Taj and its environs, and shape the forms of tourist performance.

First, the intertwined global discourses of ecology, heritage, and conservation circulate through tourist sites, focusing on specific attractions that have been assigned global importance. Indeed, the significance of a site such as the Taj has been partially disembedded from its local encoding and has become a symbol of globality (see Urry 2003a: 81). It is not merely a symbol of India now, but belongs to the world – as many commentators have noted – and accordingly is the *responsibility* of the world (see quotes from Levin and Jaffrey in Edensor 1998: 186–87). These environmental concerns are enmeshed in the growing interdependence between institutions of global governance (such as UNESCO) and national and local bodies (for instance, the Indian Supreme Court), which together instantiate new imperatives of an emergent global order in local contexts. Thus, by signing up to the World Heritage Convention, India and Agra have promoted and performed the global in the context of the local. The penalties for any failure to discharge the responsibilities inherent in the agreement could be the threat of exposure and a subsequent mediated scandal (Urry 2003a: 110–11), for these debates and controversies are increasingly subject to interventions in global media that transmit local controversies, such as the perceived environmental threats to the Taj.

Additionally, other threats constituted and discursively produced as global, notably the threat of terrorism, also circulate and centre upon the Taj. I have shown how the perceived threat of Muslim fundamentalism following the attacks of 11 September 2001, as well as heightened tension between India and Pakistan and from other internal conflicts, has produced an intensified series of security policies at and around the Taj, however unlikely an Islamic attack on a highly symbolic Muslim site might seem.

Besides these incorporating discourses and the subsequent formation of policies, the Taj has also become subject to glocal discourses and practices of expertise. I have highlighted the collaborative projects between Indian archaeologists and gardeners and landscaping experts from the United States, and I have also discussed the participation by American specialists in environmentally advantageous forms of transport in the project that 'reinvented' the cycle rickshaw. Other sharing of professional knowledge can be found in other environmental projects and within the organization of the tourist industry, which I will discuss in more detail shortly. These flows of ideas and knowledge testify to the increasingly post-national, global organization of science and technologies (Urry 2003a: 98).

However, it must not be assumed that the flows of expertise are in only one direction – from the West to India. For years, the effects of pollution on the formerly milky-white marble of the Taj have been the subject of much anxiety, and experts from far and wide have attempted to find technical solutions to restore this allure. Remarkably, despite the unsuccessful exertions of Western experts to cleanse the marble surface, an inexpensive solution was recently found in a sixteenth-century Mogul journal by an Indian archaeologist, a method that requires the building to be covered in a special mixture of mud, cereals, milk, and lime. So successful was the treatment that this mud pack is now being applied to the ancient marble structures of Rome and Florence (Syal, 2/12/2002).

Second, the debates about security, conservation, and best practice in relation to tourism become grounded in local arenas through the allegiances and appropriations adopted by local actors, testifying to the salience of the term 'glocalization'. By wielding global knowledge and discourse, local interests are able to add authority and weight to their plans and arguments. Such arguments are frequently mobilized in contests over what is or is not a suitable tourist development. For instance, in the controversy over the proposed shopping development close to the Taj, competing arguments were couched in terms of the need to further economic prosperity on the one hand, and the duty to conserve on the other. These manoeuvres articulate alliances forged between the local and the global and refute arguments that such processes simply evince a loss of local control. I have also briefly touched upon the forging of glocal lifestyles and the impact this can have on the constitution of place. For instance, Agrans of all kinds are increasingly cosmopolitan in their connections with flows of information, and particularly in the evolution of Indian consumer culture and lifestyle choices. This is shown in the adept ways that locals use to tap into cyberspace for social, cultural, or commercial reasons, and the formation of home styles that meld contemporary Western and Indian fashions with 'traditional' crafts, as in the case of the marble interiors desired by middle-class Indians.

These forms of growing awareness are resources for locals who are confronted by rapid change, as in the case of Agrans who have been hugely affected by the politicized demise of heavy industry in the city, and those in Taj Ganj who have seen the numbers of backpacker tourists on whom they depended for business severely diminish. We have seen the flexible and reflexive ways in which such local groups have responded to these sudden changes, tapping into global networks and responding to the changes in flows. The local traders in Taj Ganj have tried to diversify their tourist product, initially by searching for alternative viewpoints from which to present the Taj for the gaze of tourists unable to pay the entry fee. And the marble craftsmen have sought new markets via the internet and by capitalizing on the emerging consumer tastes identified above. The owners of factories have rapidly diversified either by moving into new areas of production or new locations. This is the reality of living in a place through which extensive global mobilities flow and continually reconstitute the locality. And out of the ubiquity of such connections emerges a reflexive competence, an awareness of information and images, that is borne out of the familiarity with the everyday effects of globalization – what Mendieta has called a 'phenomenology of globalization' (2001).

Finally, the volatility of global tourist flows, understood as species of 'ethnoscapes' (Appadurai 1990), is apparent in Agra, where there has been a severe curtailment of two formerly thick flows, namely those of backpackers and of poorer domestic tourists. In addition, fears about terrorism, which have emerged out of a melange of global and national spectres, have meant that tourist flows have begun to be intermittent and uneven. More specifically, the reconstitution of international tourist sites and their environs can be identified in terms of space and time, for flows of tourists are spatially organized according to where they stay and the routes they follow, and similarly such flows have a temporal pattern, in terms of how long tourists stay. It is of prime concern to tourist authorities to capture these flows, but as we have seen, there is vigorous competition from other nations, regions, and cities (for instance, Delhi has succeeded in capturing these flows over space and time to the disadvantage of Agra). Accordingly, it is in the interests of places to stabilize their appeal and stitch themselves into the regular itineraries of different groups of tourists. Attempts are being made to broaden the tourist appeal of Agra so that tourists can arrive there more quickly and conveniently (via the new road from Noida) and stay longer (through the development of a Taj Corridor and the linking together of other historic sites with the Taj). However, if the visiting of specific sites within these itineraries becomes problematic or unfashionable, they are likely to be bypassed, as has been the case with poorer pilgrims and backpackers at the Taj. Some tourist flows are far more lucrative than others, and so it is not surprising that the Agran and national tourist authorities are sanguine about the curtailed flow of tourists who spend relatively little, in contradistinction, for example, to package tourists. Nevertheless, the recent dearth of backpacker tourism has great consequences for the sector that depends on their business, namely the traders of Taj Ganj. This area is increasingly bypassed by tourists and is becoming marginal. This partiality reveals how the ongoing attempts to capture flows of tourists and money, often by individual actors or interest groups, can work against more general communal benefits, placing barriers in the way of greater mobility. Thus, recent developments in Agra illustrate the broader continual channelling and rechannelling of tourist flows across the world, through nations, and across local tourist spaces. Some flows expand while others thin out or are blocked. And these flows are rendered more volatile because of the effects of communications technology and mediascapes that transform local into global concerns – for instance, through the widespread reporting of the increase in entry fee at the Taj and in the response of backpackers to this development through their networks, quickened through the internet.

Places are thus continually reconstituted through the multiple flows that centre upon them, and the ways in which they are sutured into more or less stable networks of mobility, tourist and other industrial infrastructures that conjoin assemblages of other places, people, and things, creating flows along which energy, people, money, and information flow at different rates and at different scales. Places are thus dynamic and can be conjoined with networks when beneficial, or dropped as of no use when obsolete (see Edensor 2004a, 2004b). The changes in these flows, along with the adoption of glocal political imperatives, also demonstrate the ways in which local tourist space is continually transformed (examples of which include the new backpacker paths that have emerged as alternative routes from which to view the Taj; the updating of cycle rickshaws; and the emergence of new roads, and limits placed on the flow of non-ecological vehicles). What I discuss now is how tourist space in Agra is increasingly regulated, ordered, and purified. In *Tourists at the Taj*, I discussed the rise of tourist enclaves, a purified space that minimizes the impact of sensual and social intrusions on tourists through the creation of smooth surfaces, clear and smooth passages, aesthetic control, uncluttered experiences, potted versions of history, themes (Gottdiener 1997), directed and organized activity, and the general production of what might be called 'international' standards – those generic facilities, atmospheres, and forms of service found in all upmarket tourist resorts. These enclaves are constituted out of a compendium of interrelated sites – hotels, emporia, restaurants, and tourist attractions – linked by another enclavic entity, the air-conditioned, luxury tour bus. Such spaces explicitly prohibit the movement of others through them, and they extend the smooth passage for tourists between linked sites. They are also networked across the globe as spaces through which rich package tourists move, and might be considered to be species of what Marc Augé terms 'non-place' (1995). The production of green and highly regulated spaces around the Taj Mahal, the relocation of the bus parks, the de facto restriction placed on entry for poorer visitors, the blocking off of the outer court-

Sheffield Hallam University
Adsetts Centre (2)

Check-in receipt

12/06/07
02:28 pm

rism and culture : an applied perspective
ted by Erve Chambers.

5631916
ntroduction to tourism and anthropology
ter M. Burns
5643213
rism mobilities : places to play, places in

/ edited by Mimi Sheller and John Urry.
7787638

Please retain this receipt

Adsetts Centre 0114 225 2109
Collegiate LC 0114 225 2473
Psalter Lane 0114 225 2721

yard for traders and other visitors, and the building of new, plush hotels as well as the failed project to erect a nearby shopping centre all testify to further production of regulated tourist space in accordance with 'international' standards, and with the requirements of large business concerns to manage tourist flows to their own advantage. This has the effect of establishing a machine-like apparatus of policing, planning regulations, zoning policies, place-promotion, bounding spaces, and regulating flows of traffic, people, and money, to constitute a 'modulated' society through which there is a continual regulatory adjustment to conditions (Amin and Thrift 2002: 45) in order to minimize perceived disorders. This is part of a global process of spatialization that is particularly prevalent in international tourist sites

While this spatial purification seems to limit the range of performances that can be enacted in and around the Taj, curtailing the activities of backpackers and pilgrims, such spatialities must rely for their reproduction on the consent of tourists to perform in consistent and appropriate ways, and this can never be guaranteed. Moreover, a response to the prevalence of highly regulated space might be to seek out less ordered, packaged spaces, and it may be that the opening up of the global for tourists of all kinds means that new spaces evolve to serve these desires for impurity. In any case, the disorders of terrorism, pollution, and economic instability that are apt to impress themselves upon the Taj and its visitors are typical of the new global disorders that threaten tourist space, however stable and regulated it may seem.

Part 3

Remaking playful places

The paradox of a tourist centre: Hong Kong as a site of play and a place of fear

Ngai-Ling Sum with Mei-Chi So

Introduction

Economic globalization and advances in transport and communication technologies are central to increased travel. Travel by business people, tourists, and diasporas has stimulated the mobility of capital, goods, services, people, knowledge, and disease from one location to another (Urry 1995: 173; chapter 18 in this volume). Until the terrorist attacks of 11 September 2001, international travel for both play and work had taken off in an upward direction. This event adversely affected passenger numbers and miles travelled on a global scale. The SARS outbreak in spring 2003 did not have the same global impact but did severely affect travel to and from Hong Kong, some other Asian cities, and Toronto. Largely in seeking to capture these mobile flows, cities, regions, and nation-states compete to imagine and sell their historical image and locational advantages to international investors and tourists in an ever-tightening global economy (Kearns and Philo 1993; Short and Kim 1999; Hall 2000). Within the tourist industry, diverse destinations are constructed to capture the flow of tourists and to direct them to specific sites of play (Judd and Fainstein 1999). These destinations often involve the construction of various forms of 'adventure' and/or 'otherness' that can 'spice-up' the experience of specific tourist play (Hooks 1992). This chapter specifically examines Hong Kong before and after the 1997 transition from British colony to Special Administrative Region (SAR) of the People's Republic of China, and the changing nature of its East–West play within the global tourist economy.

Before 1997, Hong Kong was constructed, represented, and performed as the 'exotic' East with Western colonial characteristics. This Eastern–colonial hybrid lost some of its appeal to Western visitors with the opening up of China and of Hong Kong's reintegration into the 'motherland'. Searching for a new niche for tourists, Hong Kong is reinventing its East–West play by continuing to signify itself to the West as a version of an Eastern 'Chinese city'; but, more importantly, for visitors from the East (especially mainland China), it profiles itself as a modern tourist centre offering consumption of the West. This redirection of the tourist gaze from exotic Eastern–colonial to 'all-round' East–West simulacrum is evident in Hong Kong's announcement in November 1999 of plans to build a Disney theme park, as well as its self-presentation as a shopping site for fashionable Western goods and services for tourists, especially those from China.

While Hong Kong has been seeking to reinvent itself as a site of play, it has been subject to two paradoxical developments in the global tourist economy. These are the events of 11 September 2001 and, more seriously for Hong Kong, the emergence of

SARS in southern China eighteen months later. The speed with which SARS spread from southern China to Hong Kong and thereafter to other transport hubs with which Hong Kong has high volumes of passenger travel, such as Toronto and London, was partly related to Hong Kong's position as a global gateway in which the mobile Chinese diaspora and businesspeople engaged in extensive transnational travel. This has not only caused a setback to Hong Kong's reinvention as a tourist site for the consumption of the West, it even turned it briefly into a 'no-go area' for tourists. At the height of SARS, Hong Kong was seen as a place of fear (chapter 18). Travellers were advised not to travel there by the World Health Organization (WHO), overseas embassies, and the United States Department of State. Fear-inducing images, showing the Hong Kong population wearing face masks, were seen on the international and local media, suggesting that this site of play was overwhelmed by the fear of disease and death.

Hong Kong as a site of East–West play(s)

As a site of play, Hong Kong actively promotes itself as providing adventures in the East and West where its 'otherness' is the main attraction of the visit. This East–West play and its appeal to visitors change over time.

Before the 1997 transition, Hong Kong's 'otherness' was constructed in and through images offered by tourist brochures and advertising materials distributed by the then Hong Kong Tourist Association (HKTA) and tourist agencies. Often these brochures and the objects they constructed co-opted cultural attractions, packaging them as exotic and oriental. For example, in an HKTA tourist guide entitled *Hong Kong: British Crown Colony*, published in 1972, the literally translated meanings of the Chinese characters for two main regions of the city – Kowloon and Hong Kong – were exploited for the construction of 'otherness' (Pang 2002: 6). Hong Kong (meaning 'fragrant harbour') and Kowloon ('nine dragons') were deployed to signify 'nine dragons leaping in the fragrant harbour', a place that could offer cultural adventures and exotic play for tourists.

This exotic play was signified through oriental symbols such as Chinese junks sailing in Victoria Harbour, sticks of incense burning at Wong Tai Sin Temple, exotic Chinese cuisine in restaurants that house more than a hundred banqueting tables, dancing dragons and stilt-walkers in traditional clothing, Hakka people living on boats, as well as colourful Chinese opera in richly embroidered costumes. This sort of representation continues today – for example, the Wong Tai Sin Temple is narrated as a site dedicated to 'a hermit who had healing powers and could also make predictions about the future' in the INM Asia Guide of 2003. These symbols and images were constructed and selected to give direction to the Western gaze on Hong Kong. They were represented and packaged as objects of the 'mysterious Orient', with the promise of exotic adventure for Western travellers, at least before the bamboo curtain fell. These oriental images served as an accessible simulacrum of the East (in this case of China).

However, this particular version of the 'mysterious Orient' did not appear totally strange for Western tourists at the point of consumption. For Hong Kong was also, at that time, a British colony, connecting this Chinese space to the West. This connection, especially in the eyes of British tourists, involved nostalgia and the romanticization of empire-related desire, and regrets for a lost past. A journey to the colony promised an encounter with images of a residual Pax Britannica with the Union flag flying over Chinese soil living on 'borrowed time' and in a 'borrowed place' (Hughes 1968; Morris 1997). British colonial relics such as the statue of Queen Victoria, army barracks, naval bases, cricket ground, gentlemen's clubs, Anglican churches, and school choirs singing *and did those feet, in ancient time, walk upon* [in this case] *Hong Kong's crowded city scene*', were mixed with Chinese festivals, narrow streets and hawker stores (Cameron 1991). This created a unique East–West gaze that transgressed time and space. This transgression offered a nostalgic time-walk through Empire Lane but occurring in a Chinese cultural space aspiring to be a late-twentieth-century global city. This self-imagined romance of Empire was central to Hong Kong's East–West package and the scope it offered for a mix of self–other consumption in the global tourist economy.

As for the authenticity of this playscape (Cohen 1988), the British colonial government seldom questioned its own façade and, most of the time, the imperial symbolism of the British flag flying on Chinese soil was hardly mentioned in the official tourist literature. Instead, the latter highlighted Hong Kong's oriental identity, and its alleged authenticity as such was staged by manipulating images and experiences (MacCannell 1973). These oriental images represented a simulacrum of the East before the tearing down of the bamboo curtain of the Cold War, when China was not open to mass tourism. As far as the colonial government was concerned, Chinese traditions were promoted and preserved only to the extent that they were marketable and could be commodified without causing cultural conflicts. Those that were selected and promoted would be packaged to satisfy the Western gaze at Eastern 'otherness'.[1] This explains why neutral symbols such as Chinese fishing junks were recreated to sail in Victoria Harbour; traditional festivals were hyped for their richly varied rituals; rickshaws were placed before the Tsim Sha Tsui clock tower; and some of Hong Kong's antiquated and rickety trams were jazzed up to impart touristic meanings to visitors. These tourist icons were often no more than flamboyant tourist façades that did little to stimulate local memories or excite cultural struggles or conflicts. They were selected cultural symbols used to capture and re-fix the flows of people and money available from the mobile tourist circuit.

This play on Hong Kong's simulacrum-East and colonial-West identity was at its height in 1996–97, as Hong Kong prepared for transition from British colony into a SAR of the People's Republic of China on 30 June 1997. The 'see-it-before-it's-too-late' message brought 12 million visitors during 1996, a 15 per cent increase on the preceding year (Pang 2002: 4; see figure 11.1). Seeking to build on this success and further exploit this East–West play, the handover programme offered spectacles that involved the mixing of traditional Chinese concerts with a performance by Elton John, handover ceremonies involving national and local leaders from the West and East, most notably Tony Blair/Jiang Zemin (representing the national governments) and Chris Patten/Tung Chee Hwa (representing Hong Kong's government), and displays of oriental floats in the harbour with people watching from their yachts (Knight and Nakano 1999: 1–18). This all added to the carnivalesque nature of this historic event. Tourists flocked to take part in or witness this final episode of a distinct East–West performance; but the tourist numbers failed to surpass the 1996 record.

In the 1997 Policy Programme, the Economic Services Bureau still hoped the tourists would return, and forecast that the tourist infrastructure at the new Chek

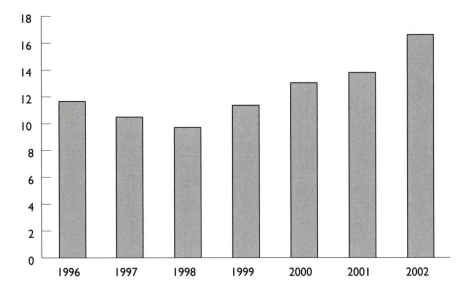

Figure 11.1   Annual visitor arrivals in Hong Kong, 1996–2002 (millions)

Lap Kok Airport should be strengthened to welcome tourists from around the world. However, after the handover, Hong Kong was beset by the 1997 Asian financial crisis, which severely affected its tourism. In 1998, visitor numbers declined sharply by 13 per cent (see figure 11.1). The absence of the expected number of tourists (and business visitors) made the new over-sized and over-capacity airport seem like a white elephant, even a mockery of Hong Kong's gung-ho attitude before the transition. This bursting of the tourist bubble since 1998 was also fuelled by other local crises such as the repeated outbreaks of bird flu and high-profile stories about tourists being fleeced in bars.

After the 1997 transition an 'all-round' East–West playscape develops. Thus the bursting of the tourist bubble forced Hong Kong to reconsider its role in tourism and imagine a new niche for tourist play (on the bursting of a similar bubble in Cyprus, see chapter 3). After its re-unification with China, Hong Kong has become just another Chinese city – but one distinctive for its identity crisis. Hong Kong can neither market its colonial imagery to the same extent nor, indeed, would it be politically advisable under Chinese sovereignty to do so. In this regard, Hong Kong has lost some of its enchantment as an East-meets-West tale, and is being challenged as an artificial replica of China when the real thing is available across the increasingly permeable border. With the opening-up and promotion of tourism to China proper, Hong Kong as its simulacrum is confronted with the 'authentic object', which is also being reinvented as a way of sustaining nationalism and state-market capitalism. Given these identity and commercial problems, Hong Kong has been forced to reimagine its time and space in order to provide a new niche of play in the global tourist circuit.

Figure 11.2   Hong Kong Harbour

In the 1998 Policy Address prepared by Tung Chee Hwa, Hong Kong SAR's first Chief Executive, a section was devoted to Hong Kong's tourism. Recognizing the success of Hong Kong's East–West play in the past, the Chief Executive noted that 'we have a unique flavour to our city, and visitors have long been attracted by our East-meets-West culture'. But he also warned that:

> Past achievements are no guarantee of success in the future[...]We must therefore take a strategic view of how best to respond to the present circumstances, and develop initiatives to maintain the interest of visitors from the mainland, from other Asian countries, and from the rest of the world.
>
> (Tung, 1998)

This call for new initiatives to rebrand the city occurs elsewhere in the world, either as part of an ongoing campaign or in connection to a particular event (Clark and Gaile 1998; Hall and Hubbard 1998; Jessop and Sum 2000). London has repackaged itself as 'the city that has everything'; Edinburgh as the 'City of Festivals' and more recently as the 'World City of Literature' (a UNESCO title it is seeking in 2005), Glasgow as the 'City of Culture' (a European Union title it held in 1990), Rotterdam as the 'City of Experience', and Singapore as 'Renaissance City'. In response to the Chief Executive's call, the HKTA in 1998 rebranded Hong Kong as 'City of Life'. This new city campaign has toned down most of the earlier orientalist discourse and is reprofiling Hong Kong as a site of play for both Western and Eastern tourists. Hong Kong continues to signify its hybrid East–West gaze, but now with different nuances. Promotional materials deploy ideas and images of Hong Kong as a 'vibrant metropolis' with a range of 'city glamour' (see figure 11.2). In an advertisement in the

London *Independent* newspaper, the Hong Kong Tourist Board narrated how a visit to Hong Kong would offer the following experience:

> It's simply dazzling! There is no better way to experience the fusion of old and new, East and West, than in Hong Kong as you cross Victoria Harbour in a traditional junk, with its spectacular backdrop of stunning skyscrapers and incredible views.
>
> (*Independent*, 14 September 2003)

This East–West spectacle is enriched by various types of 'city glamour', including designer-label clothes shops, gourmet cuisine with a fusion of East–West flavours, East–West heritage of traditional festivals, and a wild side of outlying islands and beaches (Discover Hong Kong 2003). These ideas and images construct Hong Kong as an 'all-round' East–West object of desire for all consumers, offering different adventures for each to imagine. In other words, Hong Kong becomes a smörgåsbord open to the 'pick-and-mix' and 'come-again' desires of Eastern/Western and long-haul/short-haul tourists.

Long-haul Western tourists are encouraged to imagine that Hong Kong still offers them adventures of the East – and a version that differs from those available in mainland China. In particular, Hong Kong can offer almost all of the comforts at home (modern hotels, restaurants, galleries, an Anglophone environment, and good transport and telecommunications systems). For short-haul Eastern tourists, however, especially the new middle class and *nouveaux riches* from similar ethnic backgrounds in China, Hong Kong is represented as a city that can provide 'Western' modernity and adventures not available at home (brand-name shopping, world fashion, international cuisine, financial services, cosmopolitan city landscape, and theme parks; see Pang 2002: 6–8). Thus Hong Kong as a 'City of Life' is portrayed as a modern city offering 'all-round' Eastern and Western attractions. While Westerners visit Hong Kong for its simulacrum of the East, Chinese tourists come for its simulacrum of the West.

Compared with its simulacrum East, the construction of a simulacrum West is influenced by the brand-mediated commercialization of the Pax Americana. As a growth strategy, it involves commercial co-option of brands (such as Disneyland) and the creation of pseudo-places as sites of consumption and profit (Gottdiener 1997; Hannigan 1998).[2] Regional and local élites (e.g., policymakers and policy-advising academics) play a mediating role in selecting and translating hegemonic themes/brands that are introduced and adapted to the local scene. Elsewhere, I refer to such translation and negotiation as 'sub-hegemonic', insofar as these élites are often rewarded for their capacity to translate hegemonic themes quickly into regional or local hybridized icons, identities, and interests (Sum 2003).[3] In Hong Kong, these themes include designer boutiques in shopping malls, Times Square in Causeway Bay, Ocean Park and Fisherman's Wharf in Aberdeen, Disneyland in Lantau Island, and so on. More specifically, Hong Kong Disneyland (some actually dubbed it 'Chinese Disneyland') is one of Hong Kong's latest simulacrums intended to excite the imagination of Chinese (Eastern) tourists and to refill Hong Kong's/Disney's tourist purse. Walt Disney Company (Asia Pacific) Ltd estimated that, when its new Disneyland opens in 2005, Hong Kong will gain a net economic benefit of around HK$148 billion (US$19 billion) over 40 years (Walt Disney Company 2003).

Disney is constructing, organizing, and selling the experience of Americanized pseudo-places that offer an imagined and commodified West (Zukin 1991: 217–50;

Gottdiener 1997: 68–125; Ritzer and Liska 1997: 96–109; Raz 1999: 156–91). Hong Kong Disneyland is planning four themed areas in its park (Main Street, USA; Adventureland; Fantasyland; and Tomorrowland) to lure Chinese and Eastern tourists. For example, Main Street, USA sells the American dream of a 'carefree life' around the start of the twentieth century in the following terms:

> The United States was at the crossroads of an era as the discoveries of the late 1800s were changing people's way of life. The gas lamp was being replaced by the electric light, the newfangled automobile was challenging the horse-drawn carriage for the right of way on the towns' roads and the slower pace of those times was gradually giving way to the hustle and bustle of modern life. This is the world that Walt Disney grew up in as a boy and it's the era that he looked back on fondly throughout his life as a time of carefree pleasures and happy memories.
> (Hong Kong Disneyland, *www.disney.com.hk/hkdisneyland/english/mainstreet.html*, accessed 31 January 2004)

In contrast, Tomorrowland narrates a futuristic science age in the following terms:

> Tomorrowland serves as a space oasis, a curious, light-hearted look at what life might be like on an intergalactic outpost that will take you to infinity and beyond. Here, you'll get the very real feeling that you're blasting off into outer space, while nearby you can embark on fanciful adventures with superheroes like Buzz Lightyear. As Walt Disney once said, 'Tomorrow can be a wonderful age. Scientists today are opening the doors…achievements that will benefit our children and generations to come.' But tomorrow is also for dreamers, the people with big ideas and rich imaginations. Tomorrowland at Hong Kong Disneyland has room for both in its world of fantasy and reality.
> (Hong Kong Disneyland, www.disney.com.hk/hkdisneyland/english/tomorrow-land.html, accessed 31 January 2004)

Play places such as Disneyland enable short-haul Chinese tourists to live their dreams of the modern and the future West through a mix of myths, imagination, and fantasies that relate to past or future transitions. This virtual West, found in theme parks, themed restaurants (such as the Rain Forest Café and the Hard Rock Café), and shopping malls, provide a relatively non-threatening and risk-free encounter with the fantasies and lifestyles of the West (Hannigan 1998: 71–74). In an interview with a tourist from Shanghai, a trip to Hong Kong entailed the performance of brand-named shopping not available at home. She came to join the global brand chase in Hong Kong's shopping malls. She wanted '…a Cartier watch and ring, fashionable clothes from Chanel, a Tod leather bag and cosmetics by Kose' (Cheung 2003: 1).

Because these places are non-threatening and seemingly secure, they may subtly help to bolster the pro-capitalist hegemony that is now established in China. To be successful, this hegemony must allow for certain regional and local translations, negotiations, and hybridization at the point of consumption. Seen in this context, these themed places provide Chinese tourists with a convenient and accessible taste of Western capitalist consumption. More importantly, they provide pro-hegemonic symbols and experiences that allow them to negotiate, translate, and aspire to this version of the imagined 'West' relative to China's own hoped- and planned-for transition towards an open door, high-tech, high-growth, modern economy/society that is compatible with nationalism in a period of neo-liberal globalization.

Given this move to develop Hong Kong as a simulacrum of the West, Hong Kong had to reorient its policies concerning mainland tourists. In Tung's 2001 Policy Address, mainland visitors were singled out as an answer to Hong Kong's crisis: 'The large number of increasingly affluent mainland residents constitute an important source of visitors yet to be fully tapped by our tourism industry' (2001).

Accordingly, the government announced the abolition of the quota system for the Hong Kong Group Tour Scheme,[4] which had long been used to control the influx of mainland Chinese visitors. The new targeting of mainland tourists reflects a sea change in cultural attitudes towards Chinese visitors. Once considered as 'inferior admirers' or even a 'menace' to Hong Kong's stability (being seen as illegal immigrants and money launderers), they are now regarded by the SAR as 'valuable guests'. The historical relationship between 'white guests' and 'non-white hosts' is deeply challenged by the increasing prominence of Chinese visitors (Smith 1989: 1–14).

However, as China continues to open its door and Chinese travellers become more affluent, the attractiveness of Hong Kong as a simulacrum of the West may fade. In other words, as a site of play, there is no guarantee that Hong Kong's representation of the West will not face the same fate as its exotic Eastern-colonial representation before 1997. Even before Hong Kong had a chance to test this out, it was already affected by the SARS outbreak in 2003 that led international media and tourists to regard Hong Kong as having become a place of fear.

Hong Kong as a place of fear

Hong Kong's SARS epidemic is closely connected to its complex systems of mobility as a tourist centre and, at the same time, as a global-gateway city (Sassen 2001; Sum 1999). As a global gateway, it coordinates the flows of capital, goods, services, and knowledge between the 'Greater China' region (seen as the trans-border space between Hong Kong, Taiwan, and southern China) and the global economy. This is mediated by the socio-economic activities of members of the Chinese diaspora and mobile businesspeople who travel the region and connect this trans-border space with the global economic circuit (Sum 2002). The speed with which SARS spread from southern China to Hong Kong and thereafter to 30 other countries was related to the mobility and transnational travel of these actors. According to the World Health Organization (WHO) investigation, the spread of SARS beyond China started with a professor from Guangdong – the neighbouring province in China – who travelled to Hong Kong on 21 February 2003 and infected 12 guests at the Metropole Hotel, where he had stayed. These guests were tourists and diasporic visitors and through them the virus then travelled to Toronto on 23 February, to Hanoi (Vietnam) on the 24th, to Singapore on the 25th, and so on. By 28 February, after examining the Hanoi patient, the WHO identified the infection as a new virus and named it Severe Acute Respiratory Syndrome (SARS).

On 27 March 2003, WHO recommended that all areas with recent local transmission should screen every international departing passenger to ensure that those who were sick with SARS, or in contact with SARS cases, did not travel. On 2 April, it issued a travel warning to those intending to visit Hong Kong and recommended that they postpone all but essential trips in order to minimize the international spread of SARS. These warnings began to stigmatize Hong Kong as a 'no-go area'. On 16 April, this perception was further reinforced by the United States Department of State in the following travel advice:

> This travel warning is being revised to recommend U.S. citizens consider defer-ring non-essential travel to Hong Kong because of Severe Acute Respiratory Syndrome (SARS) concerns and to inform them of updated requirements of the government of Hong Kong for anyone exhibiting SARS-like symptoms.
> (United States Department of State, www.travel.state.gov/hongkong_warning. html, accessed 31 January 2004)

Besides issuing this warning, the department also repatriated, on a voluntary basis, 'non-emergency employees' at the U.S. Consulate General in Hong Kong and all their family members. The Australian and British governments issued similar warnings to their citizens but announced no plans for repatriation. This wave of travel warnings turned Hong Kong from a 'City of Life' into a threatening pathological space of death that international travellers should avoid.

This pathologization of Hong Kong was reinforced by fear-inducing images presented on the international and local media. Newspaper reports and TV news supplied different objects of fear, starting with the daily number and mounting total of deaths and infected patients. These statistics were often interwoven with images of many Hong Kong citizens wearing face masks (for some, designer face masks), refusing hand shakes, avoiding shopping malls and restaurants like the plague, undertaking quarantine at home or in hospital/camp, passport checks being carried

Figure 11.3   Hong Kong SARS

out by masked Immigration Officers on arrival, and undergoing scanning from high-tech devices that measure body temperatures on leaving. These personal images were interspersed with social snapshots that concerned the shortage of face masks, school closures, the death of medical staff, the Rolling Stones cancelling their concerts, a joint government–WHO investigation at the Metropole Hotel, and officials in protective clothing evacuating the Amoy Garden apartment block from which 213 residents were admitted to hospital with suspected SARS (see figure 11.3). The Amoy Garden and Metropole Hotel were presented by the media as icons of fear or even death. One website even dubbed the Metropole Hotel as the 'ground zero' of SARS – making SARS, by analogy, Hong Kong's 11 September (www.nzoom.com 2003). This analogy elicited fear through its connection with the fate of the World Trade Center in New York.

This view of Hong Kong as a pathogenic place and the more general spread of SARS led to a widespread fear of travel. There was a wave of flight and hotel cancellations by tourists and business people. The Hong Kong Tourism Commission recorded a vast drop of visitor arrivals between April and June 2003 (see figure 11.4).

Cathay Pacific cancelled 42 per cent of its flights during the initial wave of panic. Their passenger levels to and from Hong Kong dropped from around 33,000 per day to 10–13,000 per day during the SARS period (Brown 2003).

But with a gradual decline of the number of deaths and new infections, the WHO announced the lifting of travel warning against Hong Kong on 23 May. The United States Department of State did likewise on 11 June. The lifting of the 'bans' signifies, at least for the local people, that Hong Kong is no longer a place of fear. The local media

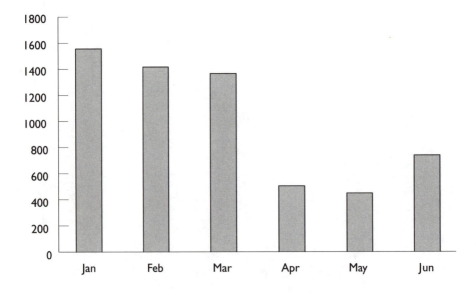

Figure 11.4   Monthly visitor arrivals in Hong Kong, 2003 (thousands)

were ecstatic with euphoric headlines such as 'Hong Kong smiles again' and 'What a difference a day makes'. The government, seeking to revive Hong Kong's image as a safe tourist and business centre, created a task force for this, which was named the Economic Relaunch Strategy Group. Some HK$1 billion (US$130 million) was earmarked for large-scale post-SARS publicity and promotion campaign to proclaim that 'Hong Kong is a safe city to visit' (Hong Kong Government 2003). Such events included visits by global football clubs such as Réal Madrid and Liverpool, and world-famous bands such as the Rolling Stones. Starting from 17 October, there was a four-week Hong Kong Harbour Fest, in which the Rolling Stones and Prince participated in promoting Hong Kong's post-SARS image. Visits by top clubs and groups who dominate the global popular culture scene may reinforce Hong Kong's Westernized image and enable it to profile itself as an 'all-round' site of play again, offering a wide variety of imagined Western consumption to Eastern tourists.

These post-SARS arrangements proved worthwhile when Guangdong travellers on group tours started to arrive for the first time after the lifting of the travel warnings on 1 June. Some were interviewed by local television and they appeared thrilled and enthused by the possibility of visiting Hong Kong's shopping malls, especially in seeing and buying the wide variety of international brand name goods. Targeting this group as the main source of visitors before the return of the Western tourists in the post-1997/post-SARS period, the government announced in August 2003 that residents from Beijing, Shanghai, Guangzhou, Shenzhen, and Zhuhai could visit Hong Kong individually – without needing to be a member of a group – from 1 September. Once all of Guangdong's 75 million residents can visit Hong Kong on this basis from 2004, it is estimated that mainland tourists alone will reach 10 million a year. As a simulacrum of the West, Hong Kong will not be short of gazers – at least until towns in southern China begin to build up their own imagined West(s), or these visitors start to venture further afield to consume an equally imagined West in the real West (as with the huge number of young Chinese consuming the West through higher education).

Conclusion

The development of Hong Kong as a tourist centre and global-gateway city reveals a paradox linked to its integration into complex mobility systems and the world of power. Regarding its role as a tourist centre, Hong Kong attracts mobile international and regional tourists in and through its continuing reinvention as an East–West site of play. Nonetheless these changing touristic meanings are all closely tied to the historical and cultural development of Hong Kong's place in the world and the changing nature of the hegemonic world order. When it was a British colony, Hong Kong represented itself as a simulacrum East with a nostalgic Western-colonial touch. A journey to the colony promised an encounter with images of a residual Pax Britannica on 'borrowed' Chinese soil. This self–other consumption for Western tourists was at its most intense as Hong Kong turned from a colony into an SAR of the People's Republic of China. This dynamic and still evolving historical transition has also prompted changes in the meaning of Hong Kong, especially its appeal as an enchanting East-meets-West tale under Pax Britannica.

Under a leadership keen to distance itself from the previous colonial era, Hong Kong has become just another Chinese city, competing head-to-head with a range of other Chinese cities, each with its own delights (such as Shanghai, Beijing, Suzhou). It thus recognizes the need to re-invent its time and space to provide a new niche of play and adventure for its present and future visitors. Influenced by the brand-mediated commercialization of Pax Americana, Hong Kong's reinvention as a 'City of Life' still signifies its East–West hybridity; but this time its 'city glamour' commercially co-opts American brands (such as Disneyland) and creates pseudo-places such as designer boutique shopping malls. This redirection of the tourist gaze in Hong Kong does not occur in a power vacuum but is shaped by changes in the global economic, political, and cultural balance of forces. The shift from Pax Britannica to Pax Americana has partly reoriented Hong Kong's East–West play. Rebranding on the Western front involves the reinvention of the 'American dream' by enabling Chinese and other regional tourists to sample its themed 'delights'. Such sampling is typically polyvalent and polysemic – involving adaptations, translations, negotiations and hybridization at the point of local consumption. It is therefore open to counter-hegemonic, sub-hegemonic, and pro-hegemonic narrations. Thus, it can be argued that when tourists claim in interviews that they can achieve 'personal self-realization' through shopping, lifestyling, fantasying, daydreaming, and pursuing a 'cool' image, they are implicitly reinforcing at the local level the hegemony of global capitalism, albeit in diverse styles. This does not imply that resistance does not occur – counter-hegemonic narrations do exist, but they lie beyond the scope of this chapter.

The rebranding of Hong Kong as a 'City of Life' project, partly under the influence of Pax Americana, has been complicated by changes linked to Hong Kong as a global-gateway city. As a gateway to the 'Greater China' region, the transnational mobilities (often including regular commuting) of the Chinese diaspora and businesspeople were important factors in the spread of SARS from southern China to Hong Kong, and then to the rest of the world. The subsequent pathologization of Hong Kong by the WHO, embassies, and the international/local media turned Hong Kong from a 'City of Life' into a place of fear. Tourists and travellers avoided this playscape 'like the plague'. As a nodal point at the confluence of the tourist and global-gateway mobility systems, Hong Kong's experience epitomizes the paradoxical juxtaposition of being a site of play *and* a place of fear. The same paradox also characterized a number of other tourist sites such as Beijing, Singapore, Bali, Mombasa, and Casablanca. These reveal the fragility and vulnerability of global spaces of play in which outbreaks of diseases such as SARS and threats of terrorism threaten the security of travellers and of their sites of play. Terrorism against tourists and the places they play in is becoming a feared and, perhaps, unavoidable part of today's tourism mobilities (see chapter 18).

Notes

1   This section concentrates on Hong Kong's East–West play from the viewpoint of Western, particularly British, tourists. With the rise of Japan and the 'Asian tigers' (Taiwan, Singapore, and South Korea), eastern tourists began to visit Hong Kong from the 1980s to gaze on a Western colony as well as to enjoy its high-quality food and buy its cheap and tax-free brand-named goods.

2   Pseudo-places are constructed for the purpose of being recognized as a familiar image (see Fussell 1980).

3   For an alternative view on the relationship between American hegemony and Chinese nationalism regarding the Disneyland project, see Lo (2001).

4   Under the Hong Kong Group Tour Scheme, mainland tourists could only come as part of organized tour groups – subject to a daily limit of 1,200 group visits.

Chapter 12

Barcelona's games: the Olympics, urban design, and global tourism

Monica Degen

Who is Barcelona?

(Barcelona City Council Campaign 1987)

B forever

(Barcelona City Council Campaign 2003)

Introduction: Amigos para siempre

Barcelona's Olympic Games closed on a hot August night in 1992 with a party in the Olympic stadium in which local residents, tourists, and sportsmen and women sang under the thundering display of breathtaking fireworks, united by the slogan: 'Amigos para siempre' (Friends forever). That night Barcelona's politicians and citizens went to bed proud of a city that had shown the world that it had shed the legacy of Franco's repression and had organized one of the most spectacular Games ever. Although there were many heroic performances, it was unanimously agreed that a major winner of the Olympics was the city of Barcelona itself, the Games not only beamed its metamorphosed urban landscape (which often featured as a background to the sporting events) into the world's gaze, but also re-asserted its Catalan pride and identity. The 1992 Olympic Games catapulted Barcelona onto a global stage and into the heart of the world's urban tourism networks. In less than 5 years the city had been transformed from a run-down industrial metropolis into one of Europe's most desirable tourist venues.

Nowadays, Barcelona is hailed as the most successful global model for post-industrial urban regeneration based on its urban design, and celebrated as a textbook example of how to turn a city into a global player. The fact that financial success or an improved global image do not always accrue to Olympic host cities is painfully illustrated through less successful Olympic ventures. Montreal, for example, is still paying off its debts for the 1976 Games, while in 1996 Atlanta almost disappeared under the weight of corporate advertising. Not surprisingly, therefore, Barcelona's triumphant reinvention has almost become a brand in itself, a model that is analysed, marketed, and followed by cities and planners around the world.

However, Barcelona's success was and is based on carefully planned strategies by a variety of players and, like every game, it has its winners, losers, and those who do not play by the rules. This chapter will argue that Barcelona's spatial transformation and development as a tourist destination are inextricably linked, and involve a complex network of local and global forces that continuously reshape the city's environment. Barcelona's success, I will argue, is based on its view of tourism as a process, rather than a product – a process that is constantly reconfigured through the flow of people, images, cultures, and objects (Rojek and Urry 1997). These flows can be captured momentarily in the actual experience of place. This experience is configured as much through the expectations the tourist brings with him/her (informed through marketing campaigns but also tourist brochures, city guides, and so on) as through the materiality of the place itself, and the activities and practices in places. It is the relationship between these dimensions on which this chapter will focus.

First, I will attempt to show how the 'Barcelona experience' is created through an intricate interplay of urban restructuring, tourism policies, and practices by both tourists and so-called locals. And, second, I will argue that as Barcelona's performance as a tourist space developed over time, the rules guiding this game changed. I begin with a historical policy overview of the city's urban and tourism development before 1992 and then discuss the impact the Olympics had on the city. While using events to reshape its urban landscape is nothing new for Barcelona, what was new about the run up to the Olympic Games is that the remake of the city grew out of a distinctively democratic impetus. Whether this is still the case in the post-Olympic developments will be evaluated in the last part of the chapter, which focuses on two areas developed since 1992 and analyses the spatial relations of these places to highlight tensions that arise from the interaction of global and local features.

Training for the 'Barcelona experience'

Barcelona has had a long history of showcasing the city through international events and using these to transform different parts of the city. The Universal Exhibition of 1888 changed the east side of the city and created the Ciutadella park. In 1929 Barcelona hosted the International Exhibition which took place on the west side of the city and saw the transformation of the Montjuic mountain area as well as the development of the underground system, and during the 1982 World Cup in Spain the city started to develop its sporting facilities. However, so far the city had not really been considered as a desirable tourist destination. In fact, Barcelona's development as a tourist venue is closely linked to Spain's political history, and the country's transition from dictatorship to democracy. During Franco's era public investment within Barcelona's city centre had been deliberately neglected as a way of punishing a city that had been the bastion of the republican and anarchist movements during the Civil War. While in the 1960s low prices and the guarantee of 'sun and sand' led to a spectacular growth of coastal tourism,[1] it was based on a policy that had been designed 'to offer the visiting foreigner images and representations of a "Spain" that was unitary, authoritarian, mono-cultural and monolingual' (Pi-Sunyer 1996: 242). There were no policies to diversify the tourist product, and cities such as Barcelona were not allowed to market and foster their own regional identity. Hence tourists just visited Barcelona as another stop on a programmed tour to see a limited number of key monuments but with no overnight stay.

The lack of interest by outsiders in Barcelona is hardly surprising as, at the time of Franco's death in 1975, the city was faced by a neglected centre, dominated by the car, regulated by chaotic planning policies that supported private developments, and lacking basic public facilities such as squares and parks. In the late 1970s much of the city of Barcelona was certainly densely populated, but in old, neglected, and degraded housing. Thus, one of the first tasks of the new democratic government was to reverse what had been termed Barcelona's 'grey period', promote it globally as the second city of the country, and resolve locally the deficits of infrastructure, public institutions, and civic facilities (Carreras i Verdaguer 1993; Subiros 1999).

Design was consciously taken up by the socialist city council (elected in 1978) in their urban development policies as a drive to reverse the disastrous policies of the previous regime. Thus, the 1980s saw the emergence of design as an expression of the new freedom, the new Catalan identity and a new Catalan style (personified through designers such as Javier Mariscal and Oscar Tusquets). Barcelona had to be physically reconstructed to symbolize through its altered cityscape the new democratic Catalonia. This transition (1979–94) was led by a team of highly educated technocrats – some of them, like the socialist mayor Pasqual Maragall, educated in the United States, and importing ideas and concepts from North American cities such as the development of city marketing strategies or the fostering of urban tourism.

Hand in hand with the rise of design, 'the street' became the metaphor and symbol for the reclamation of public life and the implementation of civic ideals in the new democratic society (Julier 1996; McNeill 1999). Indeed, as Subiros (1999) explains, Barcelona's policies for urban renewal were developed in conjunction with cultural policies. Hence, Barcelona's spatial transformations were part of a conscious cultural reinvention of a city that involved the reorganization of its time and memory. This was literally achieved by cleaning and restoring the city's architectural heritage, recovering significant places for Catalan culture, and creating collective codes of signification in the public realm. This 'Reconstruction of Barcelona' (as it was officially labelled) in democratic times, was led by the architect Oriol Bohigas, who 'rather than devising some vast masterplan...concentrated on small urban interventions – new parks and public places – that would invigorate the city's various neighbourhoods...and weave together its outlying districts with its historic centre' (Bartolucci 1996: 64). The 'Urban Spaces Project' was characterized by an emphasis on design features in public places, fostering civic spaces in the city, and a provision of infrastructures that would support new communication technologies, thus assuring Barcelona's attractiveness to international businesses (Tello i Robira 1993). The first years of planning involved extensive participation by neighbourhood associations in the creation of a new urban politics. Barcelona's new physical makeover signified its transition to a democratic modernity. As Julier remarks: 'The redesigning of public life from bus shelters, signs, and urban spaces to bars and restaurants was to communicate regional and national renewal' (1996:126).

We can see from the above discussion how from the start there was a close relationship between the regeneration of the city centre, the new democratic ideals, and developing Barcelona's new image for tourism (García and Claver 2003). Barcelona's rebirth hence was directed towards both an internal improvement of the city for its residents, and an external reappraisal of its urban imagery. What has been called the 'Barcelona model' – in terms of post-industrial reinvention and physical renewal of the urban texture – is nowadays highly acclaimed by planners and city politicians around the world, leading the city to win several prestigious architectural prizes.[2] Yet, as I have argued, this 'model' was born out of a particular set of historical and political circumstances and it is precisely the successful co-ordination of economic and political aims with strong civic commitment and participation, which was strongest during its first years, that defines this 'Barcelona model' (Marshall 2000; García and Claver 2003). The discussion also makes clear that places do not become tourist sites overnight but involve a complex assemblage of elements – of physical space, imagery, and infrastructure – as I discuss in the next section.

Setting the rules: the Olympic Games

The next decisive moment for Barcelona's city planning politics was its selection in 1986 as the host city for the 1992 summer Olympic Games. This global event would provide a unique opportunity to change the world's perception of the Catalan capital. At the time the mayor, Pasqual Maragall, openly conceded that 'we have used the Games as a pretext' to obtain three key post-Olympic objectives: improving the quality of life in the city, exploiting the economic impetus generated by the Olympic Games, and establishing Barcelona as a major European city (Barke and Towner 1996). We can see here how global processes, rather than diminishing the role of the local, express themselves in new local arenas and, in Barcelona's case, are used and reinterpreted to the city's advantage. Cities rework and situate globalization. While we have seen that Barcelona had already embarked on its physical renewal, the Olympic Games provided the city with an unprecedented springboard to become part of the 'global city' circuit.

Roche (2000) argues that to become a 'global city' certain physical characteristics need to be developed. For example, the upgrading or construction of technology and communication structures that allow the global transmission of events, transport infrastructures that allow access to a global public, and the construction of stadia and other venues that provide the setting for global events and gatherings. Barcelona's politicians and planners saw in the Olympics the opportunity to finance large-scale public-works projects through public and international investment[3] (for a detailed discussion, see García 1993), and to deliver these within a set time period. These projects included 'opening' Barcelona to the Mediterranean through a new waterfront, expanding Barcelona's grid-like structure to the sea and creating a new residential area (the Olympic Village), two new skyline communication towers, a major ring-road, a new international airport, and the refurbishment and construction of new sports stadia.[4] By reassembling the city, different elements were put in place to foster international access and visibility. It is important to point out here, as García (1993) notes, that in order to achieve local support for the event, urban transformation was debated in public and continually modified as a result of this feedback. This generated a sense of ownership and civic pride among Barcelona's residents, who had to endure a city turned into a construction site for several years.

The Olympic Games have to be understood as part of the new cultural and urban planning processes that grew in parallel with what has been termed the rise of the symbolic economy in cities. As Western cities have gradually lost their industrial base in late modernity, cultural activities and the related service economy are shaping new economic activities. 'Symbolic economy' refers to the ways in which culture as a system for producing symbols is used as a means of providing the city with an identity that makes it attractive for consumption (Zukin 1995). Increasingly, culture serves as the 'comparative advantage' in a global competition for investment (Bianchini and Parkinson 1993; Kearns and Philo 1993). The consequence for the urban physical environment is that urban areas change from landscapes of production into landscapes of consumption, so that their aesthetic appeal becomes pivotal (Zukin 1991; Urry 1995; Hannigan 1998). As Fainstein and Judd (1999) point out, Western cities are less and less able to opt out of the tourism industry, as it has infiltrated so many of their services and economic strategies. The urban tourism industry not only promotes the regular flow of tourists to a place, but also promotes the establishment of fairs, conventions, and business settlements. Yet at the same time urban tourism is an unpredictable economy: 'The three elements of urban tourism – the tourist, the tourism industry, and cities – interact to produce a complex ecological system. The tastes and desires of tourists are fickle; just like car buyers, they will yearn for next year's model even before it appears' (Fainstein and Judd 1999: 5). Hence, a city such as Barcelona that has been catapulted onto the global stage has to keep up its place by staging events that are able to attract a continuous global audience. Thus, the city council designated 2002 as the 'Year of Gaudí', which saw many of Antoni Gaudí's key buildings – which are a key element in the city's tourist imagery – open to the public for the first time and led to a 30 per cent increase in tourism.[5]

Barcelona followed this up by declaring 2003 its Year of Sport, hosting the Swimming World Championships, and for 2004 the Universal Forum of Cultures is planned. The latter is an event conceived by the Barcelona council that consists of a mixture of an international exposition and a United Nations conference where themes such as peace, multiculturalism, and freedom will be discussed. It is envisaged as being a means to finish the transformation of a working-class area in the west of the city by extending the beach area, developing the city's second-largest park on a former landfill, and thereby creating an attraction for international tourism. Affected by these developments are some of Barcelona's most vulnerable neighbourhoods, which, in contrast to developments in the 1980s, have had no say in the planning process. The Forum 2004 boldly links urban transformations featuring designed flagship environments with tourism development as recommended in Barcelona's strategic plans. These strategic plans co-ordinate key sectors of the city's economy and policy-making, and provide a forum to debate the model that Barcelona should take in the European arena (García 2003). While the first strategic plan (1990) aimed at consolidating the city as a European metropolis, and was the most socially inclusive by incorporating guidance for social redistribution, the second (1994) and third (1999) are increasingly geared towards making Barcelona a key player in a new global economy (see García 2003 for a detailed discussion). Tourism has featured throughout all three strategies, but is given a priority in the last plan, as it is part of attracting foreign private investment into the city. This has had serious consequences for Barcelona's once inclusive planning model, as García and Claver highlight: 'The new emphasis favours private investment over communal life and weakens the general consensus over redevelopment. Increasingly, the Barcelona model is a thing of the past. At the same time, tourism has continued to grow' (2003: 9–10).

The commodification of the city has led city politicians to create marketing strategies 'to promote an area or the entire city for certain activities and in some cases to "sell" parts of the city for living, consuming and productive activities' (Smyth 1994: 2). This, however, is not a straightforward process. Cities or parts of the city differ in the qualities that make them easy or difficult to sell, so some must 'undergo not only a change of image but a facelift if they are to be competitive' (Fainstein and Judd 1999: 11). Hence, different areas of Barcelona have experienced various degrees of urban change in order to comply with the tourist market. While much of Barcelona's bourgeois neighbourhoods, such as the Eixample (Barcelona's grid-like extension), have been restored, their exteriors sandblasted to remove decades of accumulated grime, critics have argued that Barcelona's planning model is increasingly eliminating the city's industrial and anarchic past as vast parts of working-class neighbourhoods are bulldozed and key working-class monuments transformed into shopping malls (Moreno and Montalbán 1991; Heeren 2002). The city's physical transformations have been constantly accompanied by a number of marketing campaigns featured on building sites, advertisement boards, and in newspaper adverts, and targeting locals as much as outsiders. From the inquisitive slogan: 'Who is Barcelona?' in 1987, it moved through 'Barcelona, make yourself pretty' (1986–99), to 'Barcelona more than ever' (1992–99) and lately 'B forever'. While this is not the space to analyse this phenomenon in detail, one can see in these campaigns how Barcelona's physical regeneration is linked to and symbolizes the economically revitalized and confident city.

Playing the game: Barcelona's post-Olympic development

Barcelona's urban transformation and tourism development is related to an increasing commodification of urban lifestyles and linked 'urban cultural products', such as music or art, since the mid 1980s. It is therefore not surprising that Barcelona's marketing campaigns are increasingly themed around architecture, design, and lifestyle, selling a specific version of Mediterranean lifestyle reflected in the physical transformation of the city and its new cultural practices. This urban imagery of 'Mediterranean-ness' is constituted and constantly reinvented through a fusion of myths, the built environment, and everyday practices permeating physical spaces, as much as it is by the city's social atmosphere (see also chapter 2 on 'Caribbean-ness'). The danger is, as Delgado (1992) concludes, that this unitary representation simplifies and homogenizes a socially complex city. As the 2002 *Time Out Guide* to Barcelona sums up:

> Along with the creation of the new Barcelona in bricks and mortar came a sponsored city promotion of Barcelona-as-concept, a seductive cocktail of architecture, imagination, tradition, style, nightlife and primary colours.
>
> (*Time Out Guide* 2002: 25)

The media itself acts as an important promoter of place myths and the 'coolness' that surrounds particular places. Certain cultural practices, people, and places are portrayed as unique and trend-setting as 'coolness has also become a vehicle for big business, the media and advertisers to push their way further into the wallets of young consumers' (Chatterton and Hollands 2003: 3). The media, fashion, and entertainment industry use urban lifestyles both as an advertising tool for their products and as a location for their production: it is cool to be on the streets of London, Barcelona, or New York.[6] The city has become both the locale for cultural producers as well as being shaped by them and 'consumers must travel to the point of production in order to partake of the local stock-in-trade' (Scott 2001: 17). Consequently significant portions of the city start to function as an ecology of commodified symbolic production and consumption (Urry 1995). Hence, Barcelona-as-concept, its mixture of avant-garde design, stylish architecture, and Mediterranean lifestyle (symbolized in open-air terraces, lively street-life, markets, and so on), can only be experienced by being *in situ*, by taking part in the everyday life of the city.

Yet tourists are not only consumers of signs (Rojek and Urry 1997; MacCannell 1989) but active players in an overall embodied experience. The intrinsic intangibility of the tourist product – the experience of taking part in the activities of a place by eating its food, smelling and feeling the sea, the social interactions with locals and tourist workers – make the tourist experience a highly sensuous engagement. These experiences can be fostered to a certain extent through policies, as we saw above, but at the same time, they are produced through the social interactions of different users of space such as residents, workers, tourists, and visitors. Thus, as cultural critics empha-size (Rojek and Urry 1997; Crouch 1999), tourism is not simply an economy or a set of sites but is in itself a set of cultural practices. Thus 'in the act of practising and performing tourism, both tourists and non-tourists (tourism workers, "locals", etc) construct and reconstruct social and spatial identities' (Hanna and Del Casino 2003: xx). As tourism and leisure expand and become an integral part of a city's economy, the boundaries between tourists and locals become increasingly blurred.

Contemporary tourism is increasingly defined by its de-differentiation from other cultural spheres (Urry 1990). Going to the theatre or visiting an art gallery are as much local as tourist activities. Through the combination of transformed physical landscapes and new spatial and social practices, a new experiential milieu is produced. Barcelona's café culture or nightlife, for example, is enjoyed as much by locals as by outsiders. Local activities mix and blend with the practices of a global mobile public. However, precisely because of the blurring of boundaries and the integration of different spatial practices and expectations, negotiations and tensions around the uses and meanings of space emerge. To explain in more detail what these tensions consist of and how they are negotiated, I discuss two places in Barcelona that have experienced urban transformations geared to tourism mobilities.

One of Barcelona's most visible transformations was to open the city to the sea. This involved transforming the old port from a functional place (and one beset by social problems such as prostitution and drug dealing) to a recreational one. In the 1980s, when work started, a seven-lane road that cut the port from the rest of the city was laid underground, so that Barcelona's famous pedestrian mall Las Ramblas would be extended to the sea. The piers were retained as wide public spaces with designed landscapes. During the 1990s two key points of attraction on the different sides of the port were built. The area of recreation known as Maremagnum, with cinemas, cocktail bars, up-market restaurants, designer shops, and a sea-life aquarium, all set in the mirror walls and with a wave landscaping created by leading local architects (see figure 12.1).

Opposite to it is the World Trade Centre area with office space, a hotel, and maritime terminals for the newly emerging global up-market cruise business. When the Maremagnum complex first opened in 1995 it was immediately a success among locals, marketed in the daytime to families and attracting Barcelona's yuppie crowd at night-time with its stylish salsa bars. As time has passed, the initial attraction of the

Figure 12.1   Barcelona Maremagnum

new has faded away, new trendy night-time venues have appeared in other parts of the city, and the stylish upper classes and families have disappeared. In the daytime it is now an eerily empty space, and at night has become a 'mainstream' place for teenagers, tourists, and immigrants. By 2002, some bar owners claimed that 80 per cent of the visitors were tourists and immigrants (*El Periodico* 11/02/2002). Increasingly the once up-market restaurants are being replaced by fast-food restaurants, and the designer shops have given space to rows of tourist-oriented stores. Since 2000 the place has indeed been increasingly avoided by a lot of Barcelona's residents when it began to be patrolled by security guards armed with batons (due to the increase of 'incidents'). The worst of these was the death of an Ecuadorian immigrant who was allegedly refused entry to the Bar Caipirinha, became aggressive, and consequently was beaten up by the door staff and thrown into the sea, where he drowned. This has led Barcelona's authorities to reduce the night licences (*El Periodico* 11/02/2002). The council is currently re-evaluating the future of this area (see also Chatterton and Hollands 2003).

Another celebrated urban renewal programme has been the regeneration of large areas of Barcelona's historic city centre, the Old City. While the Old City was not directly affected by the infrastructural changes for the Games, the moral cleansing of the Ramblas and the waterfront during the Games led to a re-evaluation of the Old City in terms of its use and symbolic role. Whereas the Gothic Quarter had been an established tourist site, neighbourhoods only minutes away such as el Raval and la Ribera were 'no-go' areas where the old, marginal, but also long-settled working-class communities lived. With their dark, winding narrow streets and reputation for prostitution, drugs, and petty crime, these areas did not attract many outsiders. After the Olympics the council made it a priority to upgrade these neighbourhoods to improve the quality of life of its residents but also to open them up for tourism. New public spaces were created, streets widened, buildings cleaned up, and historic sites marketed. In el Raval – the infamous red-light district of Barcelona – whole blocks of working-class housing were demolished to give place to the Museum of Modern Art, designed by Richard Meier. A boulevard that cut right through the prostitution area of the neighbourhood leading to the new museum was also built. This museum is only one among a series of international buildings and works of public art commissioned since the early 1980s by the city to bolster its cultural recognition around the globe. While this building, and the way this phase of regeneration has been conducted, has been severely criticized by cultural critics for its lack of consultation with local communities, the museum's public spaces are

Figure 12.2 Barcelona: 'Clos, no to the ludic ghetto'

successfully used for international events, fashion shows, happenings, and parties attracting large numbers of tourists.

Nowadays la Ribera and el Raval, often described as the new hip areas of Barcelona, with designer shops and bars lining the streets, are experiencing rebellion from their long-term residents (organized by such groups as Veïns en Defensa de la Barcelona Vella and Forum Veïnal de La Ribera). As the areas become more attractive, new residents and new spatial patterns are emerging with young people moving in, more flats rent out for tourists, new designer bars opening. Rising prices in these areas are making it difficult for locals to rent. At the same time, these spaces are also the areas where much of the new immigration is settling, often in the still-dilapidated pockets of the neighbourhood. Shops are changing and in la Ribera (Barcelona's 'SoHo'), greengrocers are giving space to boutiques and bars, making some residents question whether these are neighbourhoods to live in or just to walk around (Degen 2001). One wonders whether the needs of outsiders (tourists as much as commuters) are consciously prioritized in urban policies over those of local residents. García and Claver (2003) conclude: '[T]he city of Barcelona is becoming a home for the upper and middle classes, and its Old Town is developing some of the classic inner-city problems associated with immigration and slow-paced gentrification' (15). In la Ribera local residents' groups have come together to oppose the saturation of the area by bars that promote the 'Mediterranean' lifestyle. All over the neighbourhood, banners have been put up on balconies addressing Barcelona's current mayor Joan Clos and stating: 'No to the ludic ghetto' (see figure 12.2) or 'We have the right to sleep' and in extreme cases neighbours have thrown eggs at their night-time disturbers. Residents largely criticize the council's laissez-faire policy in granting bar licences. As a recent newspaper claims, the leisure culture is now incompatible with the residential needs of an area.

These examples raise a number of issues around the tensions that exist and negotiations that go on between the local and global interactions that emerge in urban tourist sites. For a start, they illustrate how tourism reshapes the spatial ecology of cities. While the areas have been transformed to serve Barcelona's citizens, they were also renovated to make these places desirable places to visit, and have become part of the city's tourism circuit. However, while city planners and politicians have tried to regulate these sites and foster certain forms of behaviour, these settings are unlikely to be fixed and uncontested (Degen 2003). In fact, I have argued that both tourists and residents have the potential to contest and disrupt hegemonic meanings. The Maremagnum case shows the inherent 'openness' and unpredictability of urban spaces. While the place had been designed to foster certain rites of experience, these are ephemeral and change with the taste of users and in the context of other places in the city that may offer competing experiences. The reputations that different places have – such as whether they are 'cool' or dangerous – define the relationships that people have with them and their spatial uses. These myths are interacting both on a global and local scale and their power in defining the practices in and around places should not be underestimated. As Till (2003) comments: 'A significant part of the "reality" of cities, and of places more generally, includes dreams, desires, stories and ghosts – those urban imaginaries – created by people who live and work in a place. Cities, in other words, are always "normal" and "somewhere else" for both residents and visitors' (2003: 58).

Secondly, the regeneration of el Raval and la Ribera, and their integration into the tourist circuit, highlight interesting points around the sharing of public spaces. While these neighbourhoods have undoubtedly benefited economically and socially from their touristification, those most vulnerable do not seem to have been sufficiently protected by local policies from the gentrifying features that often accompany such processes. Despite opening up these areas to a varied public, policies have not been devised that support and carefully manage a variety of attachments and patterns of spatial uses. As the needs of locals become secondary, tensions emerge between different social groups, as in the case of la Ribera. Yet it is precisely the existence of these marginal and working-class areas, their ingrained history in the textures of the place, and their social life that are the roots of the bohemian and hip environment that attracts the new cultural industries and outsiders (Zukin 1991; Gdaniec 2000). While Barcelona's regeneration initially grew out of a social-democratic project and local consensus, its success in entering the global tourist circuit means that economic concerns are beginning to overpower local needs. This is one of the consequences of becoming part of a 'global cities network', and the increased competition between locations where local policies are designed to satisfy a global market. Barcelona has been viewed as one of the cities that has managed this move most successfully: 'In many ways Barcelona has epitomized the possibility of what "third way" cities might look like – striving to be economically competitive, yet redistributing surplus in ways beneficial to a broader public, in a way that has caught the imagination of many on the left' (McNeill 2003: 75). Yet my discussion has highlighted that if there had ever been a 'Barcelona model', it is unique, and historically as well as place-specific. And, while the transformation of Barcelona's economy into a service economy reliant on tourism and events has been economically fruitful so far, relying mainly on a highly fluctuating tourist market is a risky game (García and Claver 2003).

Conclusion: friends forever?

Barcelona has indeed moved a long way since being Franco's 'forgotten city'. It has achieved celebrity status through the vibrancy and success of its place promotion, and a physical transformation that is both institutionally driven and informally communicated through tourism practices (McNeill 2003). In the course of this chapter, I have argued that much of these processes have focused on the redesign of the public realm of the city and repositioning the city's physical environment globally at the forefront of a desirable Mediterranean lifestyle. Yet becoming a player in the 'global cities network' has also meant that the rules of Barcelona's game have changed. From its initial aim to improve itself for its residents, Barcelona is now playing at appealing to a global fan base. Examining tourism as a process and highlighting the experiential qualities has shown both how global elements are embedded locally and are transformed by local forces, and at the same time how this interrelationship radically transforms everyday life in the city. By having succeeded in using the Olympic Games as a catapult into stardom, Barcelona has pushed itself into a higher league, where the stakes are higher, the game is tougher, and there is a danger of alienating its local supporters. Only time will show whether local policies can both support the internationalization of the city as well as care for and protect local needs and experiences.

Notes

1   In Catalonia tourism took hold along the Costa Brava and Costa Daurada.
2   In 1990 Barcelona was awarded the Prince of Wales prize by Harvard University for its urban reconstruction and in 1999 the Royal Institute of British Architects gave it the Royal Gold Medal for Architecture. It was the first time this prize had been allocated to a place rather than to a particular architect.
3   This period also saw the start of a culture of public–private partnerships in the economic regeneration of the city.
4   While the Games cost $8 billion (more than most Olympics have done) only 10 per cent of that total was actually spent on sports facilities (Roche 2000).
5   In fact Barcelona's tourism has increased an average of 10 per cent each year since 1994, with a record number of 3,598,201 visitors in 2002 (Patronato Municipal de Turismo 2002).
6   In a 'Cool Brand Leaders' poll in August 2003 Barcelona was chosen as second-'coolest' city after London (*www.thebrandcouncil.org*).

Chapter 13

Tourists in the concrete desert

Tue Halgreen

Introduction

A French journalist was doing research on people living on the other side of the planet, when suddenly he was struck by the following thought:

> You poor fool, you want to tell other people about other people's worlds, but you can't even be bothered to tell yourself about your own world ... What do you know about the way people live half an hour from the towers of Notre-Dame?
>
> (Maspero 1994: 7)

The rest of us may feel the same: we may have been virtually anywhere in the world, almost all places have been transformed into tourist landscapes, and with globalization spatial distinctions have become so blurred that what is geographically nearest to us sometimes seems furthest away (e.g. Harvey 1989; Rojek 1995; Urry 2002a; Zukin 1991). To the contemporary explorer, therefore, getting off the beaten track might, paradoxically, involve only half an hour's drive out of the town centre to the suburbs.

The aim of this chapter is to explore a rarely spoken of (and perhaps rarely practised) phenomenon: tourism in high-rise suburbs. The focus will be on the tourists' practices and experiences. How does such 'suburban tourism' fit in with contemporary tourism in general? Do leisurely strolls in residential areas challenge the tourist gaze, normally directed towards the extraordinary (Urry 2002a)? Or does the touristic consumption of concrete blocks show the ability of the tourist gaze to transform practically anything into an extraordinary sight?

I use the term high-rise suburb to refer to a particular type of built environment prevalent on the outskirts of many, especially Continental European, cities. The high-rise suburbs (also known as 'projects' in the United States and as 'estates' in the United Kingdom) are coherent residential developments characterized by high-rise blocks constructed from prefabricated concrete panels. They are not to be confused with the wealthy middle-class areas – made up of streets of terraced, semi-detached, or detached houses – that are more often conjured up by the word 'suburb' – though as that word implies, they share the characteristic of being located away from city centres.

Most high-rise suburbs were built in the economically optimistic decades from the mid-1950s onwards. In France, to take an example, three million council properties were built from 1955 to 1975, a period known as *les années de beton*, the concrete years. In Eastern Europe, the construction of high-rise suburbs went on until the fall of the

Iron Curtain, some estates housing more than 150,000 inhabitants (Elkins 1988). In Western Europe, the economic recession of the mid-1970s meant that, roughly speaking, workers living in these estates often found themselves unemployed, and the higher-income families moved away to be replaced by immigrants from developing countries. Today in many Western European countries, the suburbs of the concrete years are characterized by a relatively poor, multi-ethnic population, somewhat isolated from the rest of society (Rey 1996).

The high-rise suburbs are generally seen as places 'on the margin' (Shields 1991), the antithesis of the city centres they surround, yet associated with a complex and varying set of place images. They are on the one hand part of the boring and ordinary suburbia (see Silverstone 1997), and on the other hand extreme places of roughness and violence. In this chapter we see how the lack of a fixed 'place myth' makes high-rise suburbs into potential playgrounds for the visitors' own preconceptions, aversions, fascinations, and yearnings.

The chapter is based on the analysis of nine qualitative interviews and two travel accounts. The interviews were conducted in the spring of 2003. All respondents were Danes living in different parts of Copenhagen, though none of them lived in high-rise suburbs. The respondents were found through advertisements among colleagues and acquaintances.[1] All respondents had visited high-rise suburbs in leisure time, in one or more of the following cities: Prague, Bratislava, St Petersburg, Berlin, Paris, and Copenhagen. All visits (except to the suburbs of Copenhagen) had been of a few hours' duration in conjunction with 'traditional' holidays in the cities. According to the exploratory aim of the study, the respondents were asked to recount their experiences freely from a few introductory questions, such as 'where did you go?' and 'what did you do?' Two particular topics were addressed: the experience of the inhabitants and the built environment; and more practical aspects of the journey such as the decision

Figure 13.1   Brøndby Strand, SW Copenhagen

on the destination, the movements to/from and within the destination, and what it felt like to be a 'tourist' there.

In addition, I analyse two written travel accounts from the Paris suburbs. In *Roissy Express* (Maspero 1994), French author and journalist François Maspero recounts his experiences from a one-month journey through the Paris suburbs. He depicts the suburbs mostly as a place of everyday life, and the focus is on his own encounters with the inhabitants. In *Extreme Europe* (Barber 2001), British author Stephen Barber depicts the Paris suburbs as an extreme part of Europe, alongside Berlin, Marseilles, and Istanbul, which all have their own chapter in his book. Although going to some of the same places as Maspero, Barber depicts the suburbs very differently, focusing on the motorways, the pollution, the violence, and social deprivation.

Thus, this study does not intend to be representative of tourism to high-rise suburbs. It is based exclusively on the experiences of university students and university-educated people, some of which (Maspero and Barber) have a professional aim in passing on their experiences. We can only guess at the actual distribution of tourists; however, it seems reasonable to suggest that touristic visitors to high-rise suburbs are typically middle class (cf. Munt 1994a; see chapter 14 on *favela* tours). It should be stressed that tourists from other socio-economic and educational backgrounds are likely to experience such suburbs in different ways.

I start out by analysing the experience of the built environment in the high-rise suburbs. I then move on to one of the peculiar characteristics of the high-rise suburbs: the lack of life and the relation to its antithesis: excitement. Lastly, I say a little about some more practical aspects of travelling in 'non-tourist' space.

Consuming the concrete blocks

All the tourists in my survey feel the allure of the concrete blocks, although they rarely know exactly how to approach them as tourist sights. Sometimes they distance themselves from the blocks with irony and indignation, sometimes they get fascinated by their imperious aura, and sometimes they even feel an urge to touch the raw concrete. Despite the differences, the blocks are always interpreted as artefacts from a near, yet bygone era, not unlike Walter Benjamin's interpretation of the nineteenth-century arcades of Paris (see Buck-Morss 1989). In the dilapidated concrete blocks, as in the arcades, history's traces have survived, now serving as clues to the utopian dreams of the previous generation.

Thus, the tourists see the estates as reflections of grand aspirations, the belief in progress and the corresponding large-scale housing schemes of the 1960s and 1970s. Moreover, they see the estates as reflecting the aspirations of some particular people: the political leaders, the bureaucrats, and the architects. The tourists all pursue signs of technocratic planning processes, which (belying the constraints of time and cost on their original builders, as well as an informed comparison with the housing these developments replaced) they believed to be on the one hand well-meant, but on the other hand characterized by a severe lack of interest in the actual outcome. For example, the monotonous landscapes and rectilinear walking paths are signs of how the inhabitants were looked upon as objects rather than human subjects. Part of the attraction is that the high-rise suburbs are now dilapidated, with visible signs of rust or other forms of discoloration running down the sides of the blocks. The tourists see this as a reflection of the planners' lack of interest in following up their work and of the exhaustion of functionalist, utopian dreams.

Although the concrete blocks look alike, the Danish tourists interpreted them differently in the various countries. They see the blocks as signs of each country's particular political culture. In France, the large scale of the estates signifies a centralist state with a remarkable fondness for *grands projets*. In Denmark, the high-rise suburbs reflect the Social Democratic ideas of equality. And in Eastern Europe, they epitomize the utopian dreams of the brutal communist regimes (see Light 2000 on the tourist consumption of communism). Hence, in accordance with common tourist semiotics (Culler 1981), virtually the same concrete block may be seen as either 'typically French', 'typically Scandinavian', or 'typically Eastern European', depending on the geographical location (which includes the perception of the local environment as well as of the national context).

So far, the tourists agree on how to interpret the blocks. However, while some find them ugly or even appalling, others describe the blocks as fascinating or impressive. The two aesthetic judgements reflect two different attitudes towards high-rise suburbs in general. Those who do not like the sight of the blocks tend to interpret these suburbs in the light of their historical outcome. They notice how the planning schemes, whether well-meant or the result of cynical calculation, resulted in inhuman living conditions for the inhabitants. Thus, they read the buildings not only as signs of modernist architecture, but as signs of the social and cultural history of modernist architecture. Occasionally, they even read the buildings as signs of what they see as the tragic history of modernism as such – of the disappointed utopian dreams and the unfortunate consequences of a blind faith in progress.

Those who describe the concrete blocks in positive terms express their views in more aesthetic terms. The blocks are neither good nor bad, but simply fascinating for their grand dimensions, bare concrete walls, and clear signs of decay. Two respondents share this fascination, and both stress that it is a sincere feeling and not some ironic, cool attitude. They like to think of the high-rise suburbs as the implementation of visionary architects' dreams. In particular, they like to imagine how the future inhabitants were treated as anonymous pawns in the game, all believed to have exactly the same needs. The brutal appearance of the concrete blocks is believed to reflect the brutality of the planning process, and the more brutal the better. As the tourists say:

> I was thinking about the way it had all been built very quickly as prefabs by some cranes. And they've been thinking: 'This is where we want them to live, and this is where we want them to spend their leisure time.'
>
> (student of rhetoric, aged 26)

> I find the Eastern European way of thinking extremely fascinating. They build thoroughly planned flows in people's lives. They think: 'Where do we want the workers to live? How do we want them to be taken to their workplace…?' And then they just tear down at their discretion. They demolished one third of the old part of Bratislava to build a bridge that leads to the high-rise suburbs. I am very fascinated by that way of thinking.
>
> (student of computer-science, aged 30)

With its neglect of the living conditions of the inhabitants, we may interpret this experience as a case of postmodern cynicism (see Ziehe 1994). At the same time, it contains a clear distancing from the more 'soft-hearted' aspects of postmodernity. These tourists use the high-rise suburbs as the object of a nostalgic longing for a bygone era in which, as they see it, societies believed in grand theories, made forward-looking planning, and had the courage to implement such plans. In other words, this is a postmodern nostalgia for a particular modernist tendency (such as Berman 1988; Rojek 1995, on Modernity 1). Part of this experience is the fascination with what the tourists imagine to have been the urban planners' unrestricted power over people. This fascination recurs in many interviews as well as in the written accounts, most notably in the frequent use of imagined 'quotes' from the planners' thoughts, as exemplified above.

In addition, there are two more ways of consuming concrete blocks as tourist sights. In the lifestyle magazine *Wallpaper*, concrete blocks – and virtually any other sort of concrete buildings – are celebrated as extreme cases of a particular aesthetic style. For example, a concrete block in Australia (in the September 2002 issue) and a grandiose concrete stadium in a former Soviet republic (October 2001) are depicted as 'cool' tourist sights for the taste-making classes because of their aesthetic resemblance to trendy minimalist furniture. With a lot more self-distance and irony, photographer Martin Parr's collections of *Boring Postcards* turn the concrete blocks into sweet, toy-like objects by exhibiting what seems to be the childlike naivety of the 1960s utopian architecture (Parr 1999, 2000, and 2001 in particular). Thus, the collections of coloured postcards from bygone decades contain a different nostalgia than the one outlined above – a nostalgia for a simpler and more carefree past.

A thin line between boredom and excitement

Several respondents describe high-rise suburbs as 'dead' places. The concrete blocks all look the same, the shops are closed down in the middle of the day, the playgrounds are empty, and the occasional inhabitants look bored as they walk to and from the parking lots. Similarly, not a single day passed in Maspero's wanderings without him noticing a deserted residential area or a shopping mall where people go just to kill time. The youngsters are roaming around the mall and the old people 'come every day when once a week would do', as the author puts it (1994: 159). Suburban life is so ordinary that it becomes extraordinary in the eyes of the tourists. As much as the suburbanites may look bored, they never look boring.

There are three explanations of this attraction of boredom. First, boredom is such a significant part of the place myth of high-rise suburbs and other suburban environments that it may act as a particular sign of the 'typical suburb' (Urry 2002a: 12–13). In British pop and rock music, to take an example, suburban life is often depicted as a 'no life' (Frith 1997). A similar picture is painted in most other popular representations: suburbia is the epitome of ordinariness, rhythms and routines, everyday life, the family, comfort, housework, commuting, and so on. There is a pleasure of recognition in experiencing such signs of 'authentic' suburban life.

Second, bored people may act as an exotic or even desirable sight to those tourists who do not have the time to be bored themselves, either at home or on holiday.

Boredom is the extraordinary antithesis to a metropolitan life in which something *has* to happen all the time, a life characterized by an endless row of detached sensations (Bauman 1995:111–20; Simmel 1950). Moreover, with high-rise suburbs and other lower-class areas, boredom is linked to the relatively smaller life chances available to the inhabitants (Rojek 1995:156). Living in the concrete desert is not seen as a choice, but as a condition. As one respondent puts it: 'I hope they [the inhabitants] have got used to living there. After all, the estates are a condition they are bound to live with' (doctor, aged 49).

Ironic as it may sound, the tourists may for a while find this lack of options desirable. In so far as the ordinary, uneventful life in the concrete desert is seen as a social destiny, it incarnates the modern desire to escape from individual freedom (Berman 1988: 10), a desire possibly reinforced by the increased sense of contingency in late-modernity (Beck 1992).

Third, boredom is fascinating because of its inseparable link with its antithesis: excitement. From social science, political rhetoric, and popular culture, boredom and excitement are known to be causally related, with boredom leading to a quest for excitement. The boredom of suburban teenagers is imagined to lead to excessive sexuality, tragic suicides, drug addiction, or rebellion against their parents' generation (Frith 1997). Perhaps the most well-known example from the high-rise suburbs is the 1980s biography and film about teenage prostitute Christiane F., whose experimenting with drugs is depicted as the inevitable consequence of growing up in the bleak Gropiusstadt estate on the outskirts of West Berlin (Hermann and Rieck 1978; Edel 1981).

The tourists are well aware of this causal relation. Watching some children roaming around the streets, Maspero sets out the social logic of the youngest suburbanites: 'You start stealing when you're ten, when you've got nothing to do' (1994: 173). Similarly, Barber sees the connection as he juxtaposes the sight of gathering youth gangs, epitomizing the quest for excitement, with the glare of television being a distinctive sign of boredom: 'On the constrictive strips of grass between each block, gangs of children gather to affront one another at dusk; up above, the blue glare of television sets illuminates every room' (2001: 113).

In high-rise suburbs, boredom and the quest for excitement are thought to stem not only from the usual suburban lack of activities, but also from the relatively small life chances associated with living on such an estate. The youths are facing up to unemployment and the likelihood of never being able to leave the deprived estates. In desperation, they gather in violent youth gangs, if we are to believe the representations in the media (Rey 1996). No respondent in my survey recounts witnessing examples of actual violence, yet the gangs play a significant role in several tourists' experiences of the suburbs. One respondent finds the macho attitude of the immigrant gangs sexually attractive. He does not intend to get in physical contact with them, but likes to consume them visually. Barber also recounts witnessing potential violence. In this passage, he superbly juxtaposes the sexual advances of the youths with the machine guns of the police. The potentially violent situation is transformed into aesthetically pleasing scenery:

> At night, in La Courneuve, the core of life, inhabited by its overwhelmingly young population, is the bare community hall used for *raï* concerts and glaring eye-to-eye

encounters. Outside the hall, under acetylene streetlights, the police presence is heavy, every machine gun-toting, black-clad figure scanning the crowds for an infringement of the rigorous prohibitions that govern that extreme zone.

(Barber 2001: 106)

Witnessing violence or other horrifying events is not a new phenomenon in the history of tourism. Even in a commodified form, the practice has a long history (Phipps 1999). In the twentieth century, the popularity of places associated with death and disaster has been on the increase (Lennon and Foley 2000). Lennon and Foley use the notion of 'dark tourism' to refer to visits to death and disaster sites, ranging from the site where President Kennedy was shot to the gas chambers of the Holocaust (see also chapter 18 on 'places of death'). Rojek (1997) refers to similar attractions as 'black spots' and 'sensation sights', with the latter being the more current or even ongoing events.

However, although the basic fascination with violence might be the same, what the tourists experience in the high-rise suburbs does not fully fit these concepts. First, the recent focus on gang activities in the media has not (yet) produced particular attractions or 'sacralized sites', nor a canon of prominent events. No particular places or events have been transformed into commodities by the tourist industry. Watching suburban gangs as emblems of potential violence is a much more individualized and random activity. However, if for a moment we move from the suburbs of Continental Europe to the ghettos of Los Angeles (to which the Paris suburbs are often being compared), we witness how gang activities, drive-by shootings and racially fuelled rioting are becoming the objects of a more established touristic sign economy. The *Rough Guide* to the United States has a thematic article on 'The Gangs of Los Angeles', in which tourists are taught how to read scarves and graffiti as signs of such activities:

> You're unlikely to see much evidence of the gangs beyond the occasional blue or red scarf (the colours of the Crips and Bloods, the two largest gangs) tied around a street sign to denote 'territory' and widespread graffiti, illegible to most outsiders, denoting certain gangs by letters and symbols.
>
> (Cook *et al.* 1998: 898)

More importantly, high-rise suburbs differ from sites of dark tourism and sensation sites by posing a threat to the tourists themselves. It may take special precautions to separate yourself from potential incidents, and Maspero as well as several respondents in the survey all express a feeling of insecurity during their visits. Even in the British media, the Paris suburbs have received attention for their no-go areas, which the police and ambulance services only venture into if accompanied by special guards (Fenby 2002; Henley 2001). Similarly, recent films, including Mathieu Kassovitz's *La Haine* (*Hate*, 1995), have depicted life among teenagers in the most deprived estates, focusing on the violent confrontations between local gangs and public authorities. Curious visitors are likely to wonder fearfully whether they are in a no-go area or not (such areas are of course not signposted), and as the violence is almost always said to be racially fuelled, fear might increase if the visitor's skin colour is different from that of the majority of the inhabitants (see chapter 14 on the *favelas* of Rio as 'zones of risk').

The atmosphere

There is more to the high-rise suburbs than signs and sights, as the foregoing example indicates. Tourists' own preconceptions may evoke a feeling of insecurity, but the physical environment also has an emotional impact on them. There is a certain *je ne sais quoi* to the concrete desert, which all tourists mention in one way or another – although most of them have difficulty explaining exactly what it is. I will attempt to grasp this aspect of the tourist experience by using the concept of 'atmosphere', based upon Heidegger's phenomenology of existence. As the concept is new to tourism research, and as the discipline in general may be blamed for its neglect of emotional experiences (Löfgren 1999), I briefly outline the theoretical basis of the concept.

According to Heidegger, human beings are always relating to their surroundings in an emotional or attuned way. Like a musical instrument, our existence is always attuned in a certain way. We may be sad, happy, relaxed, or excited, but we never find ourselves in no mood at all (Heidegger 1962 [orig. 1927]; Bech 1996). These moods, however, do not belong to us as subjective interiors confronting (touching, reading, or gazing at) the world as an external object. The moods belong to existence, which to Heidegger is essentially situated in space, bodily as well as emotionally. Accordingly, life spaces and relations with other people are always attuned in a certain way. While Heidegger and Bech do not specifically deal with the materiality of space, others have used the word 'atmosphere' to identify those dimensions of tunings related to our being in particular places (Böhme 1995; further elaborated by Albertsen 1999). Anywhere we go, if we enter our own home, a prison cell, or St Marks Square in Venice, we are seized with a certain atmosphere. The same is the case when we step up from an underground station to be overwhelmed by rows of concrete blocks, as I shall return to below. Following our immediate experience of the phenomenon, Böhme and Albertsen claim that the atmosphere can be metaphorically said to exist *in between* the surroundings and ourselves. Thus, it can only be grasped by using a phenomenological ontology that cuts between the firm dichotomy between subject and object. Moreover, as we are dealing with an essentially pre-cognitive matter – the particular atmospheres can never be described objectively. Words can only be used to point out an atmosphere to someone who has experienced a similar one.

An important aspect of the suburban atmosphere stems from a simple paradox. You find yourself in a densely built environment, clearly intended to contain a large number of people, and yet it is virtually empty. On the large lawns made for playing, on the wide pavements made for walking, and on the great plazas created for the gathering of people, emptiness resides. Initially, the visitor reacts emotionally to this paradox. The lack of people evokes an intangible, uneasy feeling that something is wrong. The atmosphere seems not to emanate from the blocks themselves, nor from any human beings, but from this enigmatic void in between the visitor and the blocks. Some tourists in my survey are overwhelmed and scared by the atmosphere, especially when they find themselves all alone at night, surrounded by dark-grey concrete as far as the eye can see. They recount a strong emotional experience in which the darkness, the monotony and vastness of the landscape dissolve their sense of locality, giving them a feeling of being 'lost' in a space of emptiness. Others seek out the uneasy atmosphere intentionally. A 27-year-old student recounts going for a walk in the Paris suburbs late at night. She takes the last metro to the terminus – 'to make it more exciting', as she says with a playful twinkle in her eye. She then walks through the suburbs towards the city centre:

There was nothing but cold, brutal concrete – not a single person in sight. It really touched me being able to walk between these huge buildings and there was no life whatsoever. I imagined that if any life would appear, it would be a violent sort of life, an insane sort of life with groups of people trickling out from the underground.

Another part of the atmosphere is the sense of shock caused by the initial sight of enormous blocks scattered across a flat landscape. Here, the high-rise suburbs accomplish the touristic quest for the sublime, similar to the eighteenth-century travellers' quest for the majestic, yet terrifying upheavals of nature. As Löfgren has pointed out, the sublime was never merely about stimulating the visual sense, but also about being evoked by 'profound emotional disturbances' (1999: 27–28). In the high-rise suburbs, the disturbances are intensified with those tourists taking the underground to get to their destination. They leave the varied urban landscape behind, sit on the train for half an hour, then suddenly step up to the monotonous landscape of high-risers. The feeling repeats itself every time they turn a corner to face another dramatic sight: a parking lot or the sudden end of the built-up area. Most dramatically, one respondent recounts being in St Petersburg in an estate 'so huge, so tremendous, so misanthropic, that I almost started crying'.

Performing in the high-rise suburbs

Suburbia is in many ways a non-tourist space. It lacks both a material and a symbolic tourist infrastructure. No package holiday takes you there. No guidebook informs you of the shopping opportunities. And in many high-rise suburbs, no particular estate or building has a 'sacred' value in MacCannell's sense of a tourist attraction (1976), making it an obvious starting point for a leisurely stroll.[2] Moreover, suburbia differs from other non- or less-developed forms of tourist space by not being the object of a common tourist fantasy. Not only is the 'concrete desert' off the beaten tourist track, it is a place that many people would never *want* to visit.

Consequently, each tourist is faced with the task of turning the residential areas into tourist spaces for themselves. In the following, I outline some of the more practical aspects of this process, using Edensor's concept of tourist performance (1998, 2001a; see chapter 10 above). The concept has a particular relevance here because of its focus on how tourists construct and reproduce places as tourist destinations through their own praxis. With the lack of given roles to perform in the high-rise suburbs, any visit is likely to imply a considerable amount of reflection and improvisation. Edensor (2001a: 76–77) describes how some tourists deliberately seek out non-regulated, unconventional destinations that require improvisational performances. They revel in the contingency, the unexpected encounters, and the various sensual stimuli of such places. In the high-rise suburbs, however, tourists also see the lack of staging and guidance as an annoying obstacle in their quest for particular attractions. Photographing a concrete block can be a delightful play with tourist conventions, but it can also be an almost impossible task if you do not want to upset the inhabitants. In foreign cultures, the mere presence of a Western tourist is likely to turn even a residential area into provisional tourist space, thus making photography an ordinary activity (cf. Markwell 1997). A few miles from home, you may not even want to bring a camera. On the other hand, with tourists being almost non-existent, you may easily pass for a suburbanite, possibly getting an even closer look at the local customs than in most other places. In relation to these problems and possibilities, the travel accounts and the respondents in my survey suggest various performance strategies:

First, pretend to be local. As one of my respondents puts it:

> You try to behave naturally, as if you lived there. You walk like you have a clear direction in mind. You say something like 'let's go down that end of the estate', as if there was a supermarket. But you never stand still, gazing with yours arms crossed, saying 'look how grey that is!' or 'look at that balcony!' Neither do you take any photographs. It would seem very provocative. Those people who live there know very well that they live in a concrete desert and not at an attractive tourist site.
>
> (psychologist, aged 55)

This type of performance prevents one from taking pictures, from talking to the inhabitants, or in other ways pursuing your touristic curiosity. Still, several tourists in my survey chose it, mostly in fear of what would happen to them if they gave away their identity as middle-class tourists seemingly attracted to other people's misery.

Second, imagine you are a local. This is a radical case of pretending to be local, and indeed a very peculiar type of performance. One of my respondents likes to imagine that the high-rise suburb is an actual communist utopia, himself being an insignificant inhabitant in this 'living machine'. For a short while he transforms the suburb into a theme park through his praxis. He makes an effort not to see those things that do not fit in with his own image of an efficient working-class neighbourhood, for instance the recent attempts to humanize the bleak environment with pastel colours and trees. And as with any other theme park he likes to try out the 'amusements': he enters one of the blocks, takes the lift up and walks down the stairs, while imagining that he lives in that block.

Third, be a 'tourist'. Dress differently and bring a camera and a map. It will make the roles clear and make it less suspicious when you gaze at the inhabitants or do other 'bizarre' things such as photographing a concrete block. French author Maspero chooses this strategy. However, in order to cope with the awkwardness of being a tourist in an unconventional place, he occasionally overacts the role by 'playing at' being a tourist. He starts examining the stray bushes and flowers in a bleak car park, ironically regretting that he did not bring a plant guide. Such ironic performances can be a delightful game to the reflexive post-tourist. The car park or the dead residential streets might be the perfect settings for playfully questioning the orthodox notions about tourist roles and tourist sights (see Edensor 2001a; Feifer 1985; Urry 2002a). Still, a camera around the neck is not always enough to make you a harmless tourist in a place that most people would not dream of visiting. As one of my respondents was photographing – at a distance of several hundred yards – an estate on the outskirts of Copenhagen, three uniformed police officers suddenly interrupted her, asking her for an explanation on this 'peculiar activity'.

In the continuous creation of tourist space, the movements to, from, and within the destinations are of obvious importance. As Edensor's ethnographic research from the Taj Mahal shows, different types of tourists each have their own particular walking patterns, even at a site full of guides and signposts (1998). In the high-rise suburbs, the walking patterns similarly depend on the tourists' particular interests (such as architectural sights, ruined buildings, ordinary people, or gangs) and the various overall performance strategies as outlined above. Still, the lack of sacralized sights means that there are no obvious nodes around which to organize the journey, and consequently, all tourists in this study have the random stroll as their basic walking pattern, with the train station as their starting and ending point.

Suburbia and the city centre

Tourism in the suburbs makes clear the complexity of the relationship between the extraordinary and the ordinary, between the attractive and its opposite, the unattractive or the trivial. The categories are not only blurred, they seem to coexist in a dialectical relationship. The fact that some people see the high-rise suburbs as the dreariest places on earth and the last places they would ever visit provides such places with a certain record-breaking quality, making them a potential object for the tourist gaze. The challenge lies not in exploring a place that no tourist has ever been to before, but in visiting a place that a lot of people would not want to visit. Theoretically, nothing can escape this sight-producing dialectic: when the most ordinary concrete block achieves the status of extraordinarily ordinary, and thus worth a visit, then what used to be the second-most ordinary block will become the most ordinary, and so on.

However, the ironic fascination of the extraordinarily ordinary coexists with a much more conventional tourist ability to see almost anything as 'genuinely' extraordinary: the concrete blocks have their own fascinating history, the young suburbanites walk around in exciting gangs, and even the empty wastelands on the outskirts of the suburbs may have a built-in beauty, when seen from a tourist's perspective. Tourism in the high-rise suburbs seems to offer an affirmative answer to Urry's question of 'whether it is in fact possible to construct a post-modern tourist site around absolutely any object' (2001a: 92).

As mentioned in the introduction, the high-rise suburbs can be seen as marginal places, on the periphery of the cultural systems of space, and accordingly at the low end of the symbolic hierarchy of places (Shields 1991). Among the city centre's 'others' (other suburban environments, the country, and so on), the tourists in question experience the high-rise suburbs as the 'low-other'. The lack of status of the high-rise suburbs is reflected in the tourists' difficulties in finding an appropriate attitude towards the inhabitants. However, the general attitude is ambivalent, involving both reservation and fascination. On the negative side, the hyper-planned concrete environment contrasts with the organic and more varied urban environment, and on the positive side, it contrasts with what the tourists characterize as the small-minded and backward attitude of contemporary urban planning. Similarly, the suburbanites' everyday life may be seen as both unattractively bleak and fascinatingly secure, compared to life in the metropolis.

As the city centre changes, so does the perception of its Others. In Paris, as in many other Western European cities, the ongoing gentrification of city-centre neighbourhoods is said to transform them into 'suburbs' – in the sense of nice neighbourhoods with homogeneous middle-class populations. This transformation may lead to the idea that the outskirts are where the colourful life has gone to:

> Being Parisians, they themselves had for years watched their bustling *quartiers* being slowly transformed into museum-style shop windows…They themselves had hung on, but saw renovation force out the poorly off, old people, and young couples with their children, who all disappeared as rents rose and flats were sold…Where had the life gone? To the suburbs.
>
> (Maspero 1994: 16)

Maspero sees his trip to the Paris suburbs as a quest for the liveliness that has disappeared from his own neighbourhood in central Paris – and he finds it! He finds the strong communities that used to characterize his own street, and he enjoys the gastronomic delicacies of the suburban shops as well as the ubiquitous ethnic diversity. Thus, ironic as it may sound, city dwellers may use the suburbs as the object of a nostalgic longing for a lost urban life.

Notes

1   Four men and five women were interviewed, roughly half university students and half academics employed outside the university. All respondents had visited high-rise suburbs in leisure time, in one or more of the following cities: Prague, Bratislava, St Petersburg, Berlin, Paris, and Copenhagen. I shall state their age and educational background when quoting them.

2   There are, however, some 'sacred' concrete estates and individual concrete blocks around the world. A well-known example is Le Corbusier's Unité d'Habitation in Marseilles (built 1945–52), which is now a tourist landmark, a prime example of the idea of the 'living machine', and of course a distinctive sign of its celebrity architect. In New York, certain classic projects are now considered tourist sites insofar as they embody the architectural and social visions of modernism. Moreover, each country or large city may for a shorter or longer while have their own particularly (in)famous estates or blackspots, depending on local media coverage and cultural representations.

Chapter 14

Favela tours: indistinct and mapless representations of the real in Rio de Janeiro

Beatriz Jaguaribe and Kevin Hetherington

Introduction

Rio can lay claim to fame as one of the most celebrated spaces in the global imaginary of tourist pleasure sites. The beaches of Copacabana and Ipanema, the dramatic landscape into which a modernist city appears to have grown as abundantly as the tropical vegetation, the Art Deco *Christ the Redeemer* statue on Corcovado mountain that looks out over the city, samba dancing, night life, and especially Carnival, are all used to signify Rio as a desirable tourist place to play. As with all other cities whose images spark global recognition, Rio has accumulated its own repertoires of cultural invention and myth. Most typically, the image of the city is largely exported as a sensual and exoticized sun-drenched landscape of the body and of sexualized pleasure.

Too far from Europe and North America to be a mass tourist destination (though it is a popular destination with Portuguese and French tourists in particular), and too urbanized to have great appeal to backpacking Amazonian 'explorers', Rio is a city for the foreign tourist looking for 'edge' or cool, something away from the Mediterranean or Caribbean centres of tourist pleasure-seeking. It also appeals to those attracted by its gay subculture and nightlife, musical diversity, or by those who want an urban tourist experience with a beach holiday combined. Above all, to go there, whether during the torrid heat of Carnival in February or March or at another time, is to embrace a tourist space associated with exoticism and cultural hybridity.

The postcards that advertise Rio's tourist attractions usually confirm the clichés of tropical landscape, bronzed bodies, carnival dancing, and occasionally baroque churches and historical buildings. Although they do not feature in postcards and most guidebooks, the *favelas* (shanty towns of informal, poorly constructed shacks) are also closely associated to the image of Rio. Former tourist references to the *favelas* were oblique, since they included samba and Carnival as popular attractions that were to be experienced outside the terrain of the *favelas*. Yet now it is the direct contact with the *favelas* themselves that is advertised as the 'real thing' by the recent development of '*favela* tours'. Focusing on the case study of recently developed tourist '*favela* tours', this chapter explores the web of representations surrounding the *favelas* and how tourism interacts with the various types of spatiality encoded into the experience of the shantytowns of Rio.

Since the early 1980s this festive version of the 'marvellous city' has increasingly been challenged by the unmitigated disparity between rich and poor and the unchecked violence of the drug trade. Indeed, the spread of a culture of fear associated with this has become a part of Rio's representation and added a different dimension to its hedonistic image. As a major site of both violence and poverty, the *favelas* of Rio have come to the fore in a set of overlapping and competing representations of the city with which tourists have to engage.

Ambiguous and contradictory representations of the *favelas* have influenced the perception of urban life in Rio itself for a long time, but only recently through tourism mobilities. Exoticized for foreign eyes by its portrayal in realist aesthetics, the *favelas* are also a crucial subject of the media that churns a steady flow of images and narratives concerning them that have been commodified as spectacle. In this flow, the *favela* dwellers themselves produce their own repertoires of representation, which are often influenced by global images as well as by local identities. The tourists who venture into the *favelas* usually have some previous representational imagery to inform them, but the space that they enter into is a precarious one in various senses, challenging the possibility of a stable representation or gaze.

Representing the *favela*

The film *City of God* (*Cidade de Deus*, 2002) has brought Rio's poor, as well as its *favelas*, to the attention of the world. The film spans a period from the 1960s to the 1980s during which we witness an ever-increasing intensity of violence such that in the end violence becomes an absurd spectacle of social dissolution. *City of God*, however, enters into dialogue with a dense repertoire of representations of the *favelas* that have been around for some time. Journalists and writers of the early decades of the twentieth century offered a range of visions of the poor and the excluded with either more or less empathy, like today's film-makers.[1] The incursion into the *favela* perched on the hill became an adventure that baffled the mind and assaulted the senses by a profusion of smells and sights. The domains of included/excluded poverty represented by the *favela* were often absorbed as a vertiginous chaos, not as exemplars of the purposeful accelerated turmoil of the great rushing modern metropolis, but as a confusing bricolage of the uncivilized and unregulated materiality of whatever was to hand. The aesthetic experimentations of modernism drastically altered the strategies of representation and the metaphoric connotations of the *favelas*. Artistic modernism then re-encoded 'Victorian' condemnatory representations of the slum as an uncivilized rubbish tip that blocked the paths of modernity into its own aesthetic concerns. In the wake of modernism's new cultural sensibilities and ethnographic interests, the *favela* was depicted with different colours and made part of the invention of a new national identity and culture in Brazil.

It is in the context of the international repercussions of this modernist sensibility that the trip to the *favela* also became a feature of the journeys of notable foreign artists. While they had no organized tours to take them there, Marinetti, Le Corbusier, Blaise Cendrars, Camus, Orson Welles and many others ventured into the *favelas* as tourists and produced narratives, photographs, images, and architectural designs as responses of the imagination to this uncertain geographical entity. During a visit in 1936, Le Corbusier, after sighting the city from the chaotic viewpoint of the *favelas*, was moved to imagine and design a huge horizontal building that cut entirely through the city and would offer 'rooms with a view' for the inhabitants inside the ordered megastructure (Le Corbusier 1998). While residing in Rio, the American poet Elizabeth Bishop wrote several poems that metaphorized the *favela* and in one particular ballad, 'The Burglar of Babylon', she narrated the saga of the outlaw Micuçú as he attempted to escape from the police in the *favela* named Babylon. Braced with powerful binoculars, Bishop watched his struggle on the hill of Babylon, which was located above the building where she lived (1979). Bishop does not fail to make clear her incongruous social dilemma. Living in a high middle-class building that, nevertheless, was practically adjacent to the *favela*, Bishop, like her neighbours from their airborne verandas, watched the bandit Micuçú meet his doomed fate. Her voyeuristic eye, however, was charged with the pathos of such social inequality. Permeated by ambiguity, this gaze alternately saw the *favelas* as a picturesque source of vibrant samba or of a menacing city of scarcity.

Albert Camus, making a brief visit in 1949, found most of the local élite and intellectuals to be pompously boring and colonialized (Camus 1978). He valued his sojourn among the samba community in the *favelas* as an energizing respite that removed him from the fakery of the tropical bourgeoisie. Likewise Orson Welles, in his much publicized trip to Rio in the 1940s, also sought escape from the constraints of his assignments at the Office of Inter-American Affairs and attempted to include the popular culture of the *favelas* as part of the footage of his unfinished film on Brazil. But it was through the Frenchman Marcel Camus that the *favelas* of Rio de Janeiro were to achieve an unprecedented reputation. Camus's film *Black Orpheus* (1959) won international acclaim and depicted the *favela* as a vital, music-filled, lyrical, and mythical terrain.

Aside from the architectural modernist agenda that arose as a response to the chaotic jumble of the shantytowns, the modernist artistic adventure into the *favela* was shaped by a search for 'local colour' and a desire to explore the 'popular-primitive' as a source of vitality. Within architectural modernism, chaos and ambivalence are often ordered out of existence through grand designs. Le Corbusier could be said to fit this model. But the other responses, which turned the *favela* into an acceptable form of local authenticity, associated with a uniqueness of the popular that still retained a singular enchantment in the midst of what was perceived to be modernity's increasing standardization, was the more common response. In this sense, the modernist excursion to the *favela* relied on the appraising eye of tourism right from the beginning. It prized the unpackaged experience that has more recently come be sold also as the 'authentic' or the 'real thing' by tourist agencies.

In contemporary terms, the *favela* is acknowledged for maintaining itself despite the scarcity of goods and services, through a cultural hybridity that adapts, transforms, and upholds its ethos in the midst of adversity. It is a space in which the precariousness of urban life is transformed into the very image of endurance and self-regulation. Today NGO projects help some of the young *favela* dwellers by promoting cultural innovation – such as music, websites, tee-shirt design, and painting – which with tourists are a major component of their audience and market. Yet, as the plummeting prices of real estate located next to the *favelas* attest, the shantytowns are seen also as spaces of neglect and violence.

A space of the readymade, the *favela* is an icon, not of the once fashionable past or of the left-behind (as was the case of the European arcades, as Benjamin saw them) but rather of what might be called the 'left-behind of the future' – that is, a space that debunks the future promise of modernity. It represents the persistence of ambivalence, and an unmanaged bricolage that coexists with order. In urban and architectural terms the space of the *favela* is one of a space-time trajectory at odds with modernist linearity and order. Understanding time in the sense of *kairos* (the particularity of a moment) rather that *chronos* (the ordered flow of time), the *favela* is a space of the 'city-future', a possible glimpse into a nightmarish future where capital has freed itself from regulation, and urban governance and labour have followed. What characterizes the *favela*, yet is contradictory, is the overlap of its undirected mobility, makeshift material conditions, and speed of development with the fact that these are themselves central tropes of modernity. European modernity was supposed to be all about 'melting the solid into air' (Berman 1988) in a purposeful direction of change and progress. However, while the *favela* has moved fast it has done so without the direction that the processes of modernization dictated.

Favelas may eventually become the predominant part of Rio. Instead of the 'tristes tropiques' – the untouched peoples of Amazonia that Lévi-Strauss wrote about in his book of that title – this is the 'cheap tropiques'. Everything that Lévi-Strauss hated about Rio thrives in the *favela* – the confusion and hybridity, the mixture of cultures, the consumption of mass-produced trinkets. In the language of engineers and urban planners, the vision of a promised modernity implied a well-regulated mechanism progressing at a steady state and thereby facilitating the welfare of civilized citizens. To achieve this, the *favelas* had to be transformed and their inhabitants cleansed not only of poverty but also of their cultural traits. While there have been brutal and overt attempts to do this on the part of the state, more commonly, in order to achieve the desired result, *favela* inhabitants were treated as exotic figures of disorder, making them knowable, and thereby susceptible to ordering.

But the persistence of the *favela* and its coming to prominence as a space of uncontrolled violence has had a way of unsettling Rio's glamorous reputation, and shows the incompleteness of the civilizing project. Whereas Baron Haussmann successfully got rid of the stasis of the medieval past in Paris when he redesigned the city's boulevards as part of his modernizing project to enable things to move freely (Berman 1998), the municipal authorities of Rio de Janeiro were defeated in their similar efforts because the *favelas* surpassed their control. Not only has their overwhelming expansion refused to be contained, but the political perception of *favela* culture, and of the rights of the poor, have become officially acknowledged. From the late 1990s it became evident that the *favelas* were not only here to stay, but that they were becoming the image of the city itself. The urbanization of the *favelas*, rather than their removal, is what is now being tentatively considered.

In key respects turning *favelas* into tourist sites contributes to this process. The *favela* is recognized not just as a space but as an event, the source of cultural innovation and flow: samba in the 1930s, hip-hop and other musical styles today. The metaphoric and representational uncertainties surrounding the *favelas*, the uneasy negotiation between national popular culture and marginalized outcasts, the embrace of hybridity, and the fear of social violence are further intensified with the emergence of a new kind of visibility in the media. This new visibility offers a range of representations of the *favelas*, the poor, and the outlaws. The new ingredient is given by the internalization of the media by the drug dealers and other *favela* dwellers. Does this represent purification by direct control or purification through transforming uncertainty into an exotic reality? It seems that it is more the latter, transforming uncertainty into an exotic reality. Indeed we hesitate to include the notion of purification because the modernist cultural invention in Brazil was precisely a celebration of hybridity, of national intermingling of cultures and races. The difference in contemporary discussions around the theme of hybridity is that previously these mixings had to be subsumed to a national canon and narrative. The contribution of Brazil and other countries in Latin America is that before the term 'hybridity' became a fashionable academic and global word, it was already implied by the continuous discussions on cultural 'miscegenation'. Nonetheless such miscegenation is an ordering strategy and the *favela* has been understood in these terms.

The promise of the modern city was precisely the possibility of envisioning alternative futures distinct from the tradition-bound premises of the past, and different also from the hierarchical constraints of rural life. As either the scenario of the revolutionary masses or the conquering ground of the enterprising individual, the modern city was to fulfil the dreams and aspirations of a better future. In Rio de Janeiro, dramatic social violence such as the massacre in 1993 of *favela* dwellers in Vigário Geral by police forces seeking revenge on local drug dealers, and the murder of street children, also in 1993, in front of Rio de Janeiro's cathedral were among a host of events that spelled the carnage of unleashed violence. In the wake of these events of attempted 'purification', the term 'divided city', coined by journalist Zuenir Ventura in his book of the same title, gained currency as an apt description of Rio de Janeiro.[2] Yet such polarizations reveal that more than just a 'divided city', Rio de Janeiro is a tumultuous, fluid urban maze of inequality and social juxtaposition. Between the *favelas* and the neighbourhoods of the rich and the middle class are numerous exchanges, and indeed it is the ambiguity of these indistinct contact zones that allows violence and socialization to occur simultaneously.

The persistence of such acute social polarization reflects both the failures of modernizing projects as well as their triumph. *Favelas* have increased throughout Brazil. Yet the term *favela* encompasses such an urban variety that it no longer has a single meaning. If there are many aspects to the lives of the poor and a diversity of cultures of poverty, there are also many forms of social critique that have increasingly surfaced within the urban tissues of the great cities. The global impact of the media, neighbourhood associations, NGOs, and new role models set by the several agendas of identity politics provide a wide range of social options. The struggle for representation is particularly relevant in the depiction of the poor, marginalized, and oppressed. It is in that context that tourists who come to Rio, no doubt intending to see the famous landmarks and spend much of their time on the beach, in the historic buildings and museums of the city centre, and in the up-market shops of Ipanema, are called on to engage with the *favelas* by paying their $30 at the hotel desk and going on a jeep tour of Rocinha.

Although the *favela* has been exhaustively explored in the media, artistic expressions, and academic research, it has never before been the overt subject of the tourist industry. As mentioned above, from the 1930s until the late 1950s, illustrious voyagers to Rio ventured into the 'exotic' terrain of the *favela*. But their undertaking was an individual enterprise and not an organized or packaged one. Popular tourism did not immediately follow on from this. During the military regime from the 1960s to the late 1980s, the touristic endeavours of the municipal authorities were geared towards attenuating the visibility of the *favela*. In the most repressive period of the military dictatorship, the early 1970s, such efforts often resulted in comically crude strategies, as illustrated by the much-publicized reception given to the American tourism confederation ASTA (see chapter 12 above on Franco's Spain). In this particular event, the representatives of ASTA were driven from the airport into the city on a highway flanked by gigantic screens that depicted the Christ statue with the caption 'Rio Welcomes ASTA with Open Arms'. These screens hid from the gaze of the tourist agents the *favelas* that had grown up alongside the highway. But now the view of the *favela* has been replaced by a democratic competition for various representational recuperations of the *favela*, struggling to cover the unstable terrain of the shantytown. The previously prohibited connection between the *favela* and tourism becomes one of the new representational corridors in which the representation of Rio is being articulated.

Favela tours are now routinely sold at the hotel desk. These tours aim to mitigate urban squalor, violence, and poverty by highlighting how the organic community with its enterprising, celebratory ethos keeps abreast of its adversities. Rio can be redeemed by the *favela* only if it is understood as a reality rather than a myth – that is the message that we wish to investigate here by looking at the *favela* tours to Rocinha.[3]

Favelas and the tourist gaze

Only the very rich and the very poor live on the sides of mountains. The shantytown of Rocinha (to the west of the beach areas of Ipanema and Leblon) has spread like a compact maze layering the sides of the surrounding slopes. At night, thousands of illuminated windows transform the *favela* into a blinking constellation seemingly off in the distance when seen from the affluent southern zone. Facing Rocinha, the towers of the gated condominiums of São Conrado are symmetrically arranged amid manicured shrubbery and the rectangular, abstract blueness of numerous swimming pools. Two versions of a tropical modernity confront each other. These buildings do not adhere to the rhizomatic topography of the clinging constructions of the *favela*, but stand upright with their rising arboreal silhouettes outlined against the mountains and the messy background blur of Rocinha. When darkness falls, their lights shine in geometric patches, in contrast to the chaotic sprawl of their neighbour.

Separate as these urban forms may be, there is also an inevitable mingling. Rocinha's privileged location between wealthy neighbourhoods and the beach of São Conrado produces an intermingling of classes in the *favela* itself, on the beach of São Conrado, and to a lesser extent inside the exclusive São Conrado shopping mall. Democratic access to public spaces and the enclosed corridors of the shopping mall are threaded by financial obstacles and mechanisms of social discrimination. The São Conrado Fashion Mall avoids having cheap chain stores and features only one McDonald's that is positioned on the lower level of the mall facing the doors. Contact with their upper-middle-class and middle-class neighbours from the condominiums and houses is largely conditioned by the employment of people from Rocinha in subaltern positions such as porters, maids, caretakers, and cooks. The main hotels of the area also rely on the working population of the *favela* to form the bulk of their employees.

Often compared to an unruly labyrinth, or an organic city of improbable architecture, Rocinha is so large that it contains several subdivisions within its urban maze. Houses are packed and jammed together, built on top of each other – one owner below often selling their roof space to another above to build on. Expanding construction threatens to devour the remaining patches of green that still survive in the mountains nearby. As it spills over the other side of the hill next to the high-class neighbourhood of Gávea, Rocinha encroaches onto the edges of the green areas of the protected municipal city park.

Since 1996, Rocinha has been promoted to the bureaucratic status of a neighbourhood by the municipal authorities. Yet the dwellers of Rocinha and São Conrado consider it to be not a neighbourhood but a *favela*, and one so large that it is almost a city within the city. Comprising both residential and commercial areas, Rocinha displays considerable urban and social diversity. It has its own demarcations between the very poor, the poor, and the precarious middle-class sectors. There is a thriving housing market within the *favela*. Rather than being a point of first destination for the rural migrant arriving in the city, one needs to know someone, have connections as well as money, to be able to get in. Those without family connections more often have to start in one of the more squalid *favelas* to the north of Rio or on the streets downtown, and move up in the world to a *favela* residence in a place such as Rocinha once they have established connections and made some money.

A flâneur in Rocinha experiences its urban diversity by walking through commercial streets and alleys, and by viewing its churches, shops, bars, and stores of all kinds. Small motorbike taxis cruise up and down the hills, children, some on their way home from one of its schools, scramble by in groups, drug dealers make an occasional appearance with machine guns, tourists wave atop decommissioned army jeeps. In a matter of minutes, the wanderer can go through alleys packed with rough brick houses, streets of shacks made of planks, and unpaved lanes strewn with rubbish, dog excrement and raw sewage (especially in an area called 'Roupa Suja'), and then climb up the hill to the Laboriaux. At the tip of the *favela*, the Laboriaux commands a magnificent view of the lagoon, Ipanema beach, and Corcorvado mountain. Postcard views of Rio de Janeiro are seen daily by a handful of dwellers who live at the top, where all have rooms with a view.

With a population estimated at up to 200,000, Rocinha is the most globally connected *favela*. With 90 per cent of the homes having television, with two cable TV services, several radio stations and two newspapers, Rocinha is a mediatized *favela*. Yet it remains very much a shantytown in its scarcity of health services, schools, and sanitation.[4] It is still a *favela* in relation to the overbearing presence of the rampant drug trade and to the culture of violence that is directly linked to its expansion and repression. Yet the drug trade itself is largely accountable for the *favela*'s globalization, since it places Rocinha on the map of the transnational drug business. The community now even has its own website.[5] With a population largely made up of immigrants from the northeast, Rocinha has peculiar patterns of social organization. It is still predominantly Catholic, although the usual presence of *umbanda* (a syncretic hybrid Catholic/folk/spiritualist religion), *candomblé* (a more esoteric, Yoruban-based religion), and the evangelical churches are represented in a series of temples and places of worship dotted about in its winding streets.

Another crucial aspect of the *favela*'s global reach is centred on its tourist trade. Aside from the travel agencies that are located in the *favela* and that cater to the community, Rocinha itself is part of the '*Favela* Tour', the 'Jeep Tour', and the 'Exotic Tour', all of which bring foreign visitors to the *favela*. Aimed principally at North

American tourists, one can book online or through a hotel to go on a half-day trip into Rocinha. The jeep – army surplus bought cheap but still painted in military green – leaves from the hotels in the tourist beach area of Copacabana and travels into the *favela*. Tourists have local guides, sometimes people from the *favela* who have grown up within its streets, sometimes students using their earnings to help them through college. They have been given rudimentary training courses as tourist guides and in the etiquette of how to deal with foreigners. Although some of the *favela* tours insist on the motif of exoticized poverty and instil the sense of venturing out on a social safari by piling the tourists atop a jeep, or by directly framing their gaze by using 'exotic' as a brand name, the most commonplace assertion of interviewed tourists, and of the shop-owners who attended them, is that foreigners want to view the 'real' Brazil, the hidden picture not included in the postcards and brochures about the 'Marvellous City'. The narrative of the tour is often an NGO-inspired one of 'people coping with adversity' and 'successfully dealing with social problems'. But the frisson of danger, of the exotic, is still there, and through the mixing of these two narratives the 'real Brazil' is encountered.

A postcard[6]

> The jeep picks us up from the hotel in Copacabana. We drive about four miles to the *favela*. Our guide asks us who we are and tells us about herself – she is a student, but also a teacher who does at least two other jobs, including this. We pass by the almost fortified American school for middle-class people with aspirations for their children before arriving at Rocinha further up the hill. The jeep stops at a viewpoint from which we can see back over the city and the Christ the Redeemer Statue in the distance. There is an arts-and-crafts stall at the viewpoint. *Favela* children, trained by a local artist, sell their paintings through this stall. The guide encourages the tourist to buy something but not to give money to anyone begging. After that it is a negotiation between the guide and the tourists. For the nervous the jeep will then drive them through the *favela* to the bottom, provide a commentary of the sights on the way, and return them to their hotel. The more adventurous set off on foot and are picked up by the jeep at the end of the tour and returned to the hotel.
>
> The foot tour takes about two hours and is all downhill. A recognized guide is essential. The unaccompanied person is likely to be confronted – they might be undercover police. The tourist is taken down the street, past a municipal water house, and a municipal building that includes a library. Basic amenities are pointed out. At the bottom of the street we are taken into a house that is for sale. The owner gets R\$1 for every visitor – some can make R\$300 a week, about double what one needs to subsist in Rocinha. We see inside a house and go up onto the roof and look down over the *favela*. Children are returning from school at the bottom, most on foot, those who can afford it on the back of motorcycle taxis.
>
> The tour continues down the hill, past a police station, banks, bars, and a bus depot. At weekends hip-hop, forró, samba, and a medley of other music events can be found all over the *favela*. CDs by local musicians are on sale at a kiosk. There is no mention of the drug trade or violence on the tour. The people seem happy –

this is after all a fairly up-market *favela* compared to some in other parts of the city. We see a gym, an IT centre offering courses and internet connections, and a small, booth-style branch of McDonalds. Then another stop at a shop selling locally made crafts, coasters made from recycled paper, screen-printed tee-shirts bearing 'come into Rocinha' slogans, postcards, carved incense burners, more paintings. They take credit cards, though local currency is accepted as well as dollars. Some fire-crackers go off. 'What are they for?' For letting people know that the police are in the vicinity we are told – somewhat reluctantly. Or, as someone later tells us, more likely for letting the dealers know that a supply of drugs has just arrived. There is reticence on the part of our guide to tell us this. It is dismissed as of little consequence, 'yes there are drugs and violence but only a minority are involved'. The typical discourse here is derived from NGO ideas that these constitute local responses to social deprivation. Finally at the bottom the main market area, butchers, bakers, groceries, clothes, and more arts and crafts are available. Sights, yes, but also aromas, heat, bustle, noise. We make our third stop at a bar. There are cheese pasties for sale, caipirinha – Rio's famous cachaça and lime-juice-based cocktail – bottled water, and Coca-Cola. Then we rejoin our driver and are driven back to the hotel. Pictures have been taken, souvenirs bought, local colour experienced – a mix of what Urry calls the anthropological and spectatorial gaze (1995). A singular gaze is not afforded by such an indistinct and fluid space.

In contrast to the well-known terrain of middle-class consumption found in the modernistic buildings of São Conrado and its Fashion Mall, the foreign tourists of the *favela* tours are mostly encouraged to be interested in the cultural specificities of the socialization of the poor, and also in a certain 'shock of the real' made tangible by the poor quality of the *favelas* buildings and by the raw sewage and rubbish that overflows in many of the unpaved lanes. Given its panoramic views, relative calm, seemingly happy and welcoming inhabitants, and the proximity to the wealthy areas of the city, the tour to Rocinha can be experienced and is often packaged as an interlude, a slice of the 'real' that does not actually entail a very close-range mingling with the unglamorous poverty of a housing estate such as Cidade de Deus or the other *favelas* located in the western and northern sections of the city. This is the real as spectacle. It is a real framed by the mixing of urban deprivation with images of glamour, different in kind from the visits to the usual sites and the encounter with the standard set of tourist non-places of airport, shopping mall, and hotel. The *favela* can no longer be ignored in representations of Rio as a tourist destination, so it has become a part of that discourse. It is incorporated, but continues to resist. As an indistinct zone it retains an unsettling narrative position that calls for emphasis, and affords nervousness, on the part of the guide.

Understanding the tourist *favela*

In some ways MacCannell's description of the tourist's search for authenticity, and acceptance of its staged form, appears to hold true here (1989). The search for a back-stage to Rio (in Goffman's terms), a space that is more truthful than the artifice of the hotel, beach, or other tourist site, is what is on offer to tourists through the jeep tour.

Such is the discourse of tourism promoted through the tour: a half-day peek behind the scenes of Rio away from the 'routine' of the glamorous beaches. But it is not what is offered in practice. The *favela*'s heterogeneity of form and representation blurs the distinction between front- and backstage, and challenges the spatial configuration implied in MacCannell's concerns with the tourist use of a spatial regionalism. To move through such a space is to encounter more than one reality, and more than one form of staging, framing, or artifice. Poverty and violence, community and self-help, media cool – this space is fluid rather than regional. It does not afford a singular reading or a singular gaze. It is lived through rather than just seen, embodied as well as projected. The experience of the tourist spectacle in such a space is revealed as a kinaesthetic as well as an aesthetic one. One cannot detach oneself from this space as one moves through it. It is a slum but it is also sanitized. One is guided and therefore supposedly safe. It is part of exotic Rio but it lacks Rio's more obvious glamour – calls it into question, in fact, but without erasing it. It promotes a new version of the exotic, one different from the beaches or Carnival.

The nature of Rocinha as spectacle is not that it presents a simulation of some hidden or authentic reality, but rather a series of layerings of the real that are experienced through the tourist's movement through space. These layerings enfold surface and depth, re-ordering this binary logic through performed, negotiated experiences through the gaze, through the camera, through the safety of the jeep, through less certain movement on foot, through remembered film or television representations, through the smells and the feel of a place. The tourist's reality is therefore a condensation of all the representations of the tangible reality, producing an intangible sense of the 'performed experience' of simply being there. To gaze and to photograph is to try to capture that experience and to represent it; but the *favela* refuses easy representations.

The space of the tour is the space of mingling. The tourist mingles what they encounter on their tour with the representations of Rio they already carry with them. One can be blasé or bewildered, full of excitement or fear. All the ingredients of an action film are there in the *favela*, made visible in the contention between the lawless, ruthless drug lords and their squadrons of gun-toting adolescents and the corrupt, vicious, and vengeful policemen. In this buffer zone is the population of the *favelas* and the hardworking citizens trying to make ends meet, to raise families, and to aim towards some kind of future.

For foreign eyes, the poor of the *favelas* of Rio de Janeiro are made doubly 'attractive' not only because they are striving to overcome their daily hardships but also because their manner of surpassing obstacles is so compelling. The conscientious poor are adjusting to the languages and representations of the underprivileged that circulate through NGOs and other globalized networks, and the 'organic' poor represent local cultures with samba schools and hybrid adaptations of funk and hip-hop. Yet the attractions of the community are highlighted because postcards of the city of violence and scarcity have already circulated, and the images of horror have become part of the sun, sex, and samba scenario of Rio. Visitors want to experience this aspect of Rio's reality, and the *favela* tours become a feasible way of glimpsing this zone of risk without really taking much risk. The motivations for undertaking a *favela* tour are evidently varied, but the maintenance of the flow of tourists is only feasible because such tours deliver what they promise. The spectacle of poverty is granted and yet the sense of 'packaging' is diminished because the local people are not 'performing' for foreign eyes. Rather, it is the space that performs.

Conclusion

Favelas have resurfaced in contemporary urban imagery as a synecdoche for the contradictory nature of a national dilemma, and as the crucial locus that will decide the fate of metropolitan existence in Rio. The illicit and flourishing drug trade is largely responsible for the new visibility of the *favela* as the battle between drug dealers, police, and the state has fermented a culture of violence that surpasses in scale and devastation all previous causes of urban violence. Furthermore, the overwhelming presence of the media, centred foremost on television and the circulation of globalized consumer goods, lifestyles, and political agendas, has transformed expectations and cultural identities. Such transformations are keenly felt in the arena of youth cultures, where the former national samba has lost much of its terrain to funk and hip-hop in the *favelas* of Rio de Janeiro and São Paulo. An image of a globalized *favela* emerges as the former national portraits of Brazil become increasingly fragmented. It is into this uncertain, creative, devastating space that the tourist unwittingly enters.

The *favela* of Rocinha is indeed a formidable maze, and the impact of viewing so many people jammed together in the precarious alleys, the juxtaposition of utmost poverty with enclaves of middle-class buildings and consumer goods, and the contrast between architectural flimsiness and the breathtaking views from the top of the *favela*, are enough to give you your 'money's worth'. As the representational dispute surrounding the *favelas* thickens, the number of journalists, anthropologists, and researchers increases. Within the multiplication of images and narratives, *favela* dwellers have not only learnt to select their own repertoire of reality but have also – as has so often been documented by woeful travellers concerning 'native authenticity' – recoded their speech and mannerisms according to the ways in which they believe they are 'supposed' to speak. In other words, they have also internalized mainstream media versions of themselves, in which everyone is in their own reality show.

Viewed as both the postcard of Brazil and the turbulent terrain of unleashed violence, Rio de Janeiro is a city represented by contradictions. The complex relations between the *favelas* and the city have become a crucial arena in a contest of representations of these contradictory realities. Alternately represented as the locus of an authentic and exotic 'imagined popular community' and as the terrain of danger, social violence, and poverty, the *favelas* have become icons of the city's inability to overcome its disparities and social dilemmas. As unplanned urban agglomerations without basic infrastructure, and inhabited by people who do not hold legal ownership of the land they occupy, the *favelas* are technically speaking 'cities without maps'. We suggest the notion of a mapless city in order to produce a metaphor of the 'indistinct contact zone' between representations of the city and of the *favela*.[7] The antithesis of the Western suburb, this rhizomatic space offers a contrasting, alternate order to that of the city in general. It can only be understood at street level, at the level of performance – from the air these zones of the city look merely chaotic. That alternate order is a product of migration, capital, and the land itself. It brings forth new urban forms, cultural synergies, and challenges to the social relations of the city, while at the same time reproducing a social structure, power relations, and inequalities. A surface of emergence for new urban dynamics, cultural production, and forms of mediation, such indistinct contact zones are porous, contradictory, and permanently negotiated. They are constructed by daily social practices and imageries, producing shifting mental maps that lie outside conventional urban imageries. Those familiar with Brazil's cinematic tradition of neo-realism, most recently reworked in a new aesthetic form by *City of God*, will have a sense that they have entered such a space before. An Augustinian space this is not.

Postscript

During the Good Friday of the Easter holidays of April 2004, the drug leader of the *favela* Vidigal adjacent to the Rocinha attempted to invade it in order to oust the local drug trafficker. A major shootout between the rival drug gangs burst out and when the police attempted to intervene, the entire areas of the Rocinha, Vidigal and the neighborhoods of São Conrado and Gávea were practically under siege. Twelve innocent people were caught in the crossfire and died. The irruption of such violence in the Rocinha impacts directly on the lives of those living in the wealthy neighborhood next to the *favela* and on the entire circulation of traffic of the southern zone of Rio de Janeiro. Major avenues were shut down, public schools next to the *favela* and the extremely prosperous American School in Gávea near the Rocinha were also closed down. The shootout galvanized public opinion and undermined the generalized view that Rocinha was progressively becoming a more peaceful and prosperous community. Due to its strategic location and its relative wealth and proximity to prosperous neighborhoods, the drug trade in the Rocinha is the most profitable of Rio de Janeiro. Once again, NGO efforts, neighborhood associations and the *favelas* tour itself are held at checkpoint. As the most mediatic *favela*, the Rocinha also received enormous newspaper and television coverage. Violence and despair in the community produced a 'shock of the real'. Yet, violence also cancelled representational efforts, symbolic narratives, and political agency. After the shootout, the *favela* attempts to regain its life and to live beyond the stray bullet.

Notes

1   For a discussion of the modernization policies of Brazilian engineers at the beginning of the twentieth century see De Carvalho 1994.
2   Ventura 1994. For a discussion of the concept of the 'divided city' and the press coverage of the *favela*, see Cavalcanti 2001.
3   Aside from the small community in São Conrado, Rocinha (formerly known as the largest *favela* in Latin America) is the only *favela* so far that has organized tours. In this chapter, we focus on the particular case study of the *favela* tours organized by local agents in Rio de Janeiro.
4   Rocinha has an estimated area of 722 sq km. Television ownership statistics are according to data seen on *Jornal do Brasil* in 1998. The size of Rocinha's population fluctuates tremendously. The research institute IBGE estimates 56,000 people, whereas the neighbourhood associations claim 200,000. *Istoé* 6/5/98 estimates the figure as 150,000. Statistics from Luiz Cesar de Queiroz Ribeiro from 'Habitação' (Ippur: UFRJ) show that in the last half century the population of Rio increased overall by 134 per cent (from 2,375,000 to 5,551,000) though the *favelas* taken alone have expanded 463 per cent. Charts from de Oliveira 2001 and the IBGE give the following list of infrastructure: 2 radio stations, 3 schools, 1 samba school, 2 showrooms, 8 taxi ranks, 4 pastry shops, 2 supermarkets, 10 bakeries, 20 butchers, 14 drugstores, 20 evangelical churches, 4 bus lines, 2 lottery houses, 2 local newspapers, 10 nursery schools, 1 public health outpost and 1 private health post, 1 refuse collection centre, 1 water supply station, 2 police posts, 1 federal bank branch, 1 state bank branch, 2 TV stations, 2 women's associations, 3 neighbourhood associations, and several NGOs.
5   See *www.rocinha.com.br*
6   This description is based on a tour undertaken in September 2003.
7   For a specific interpretation of the notion of 'contact zone', see Pratt 1992.

Part 4

New playful places

Chapter 15

Playing online and between the lines: round-the-world websites as virtual places to play

Jennie Germann Molz

Introduction

Since mid-2001, the online travel company Orbitz has been running an advertising campaign with colourful avant-garde graphics depicting various destinations around the world. The advertisements bear the catch phrase 'Visit Planet Earth'. This inviting slogan raises an interesting question: how can the global become a tourist destination? In other words, in addition to thinking about how particular places to play become global, can we think about how the global itself becomes a place to play?

The notion of the world as a place to play began to emerge in the late nineteenth century when Thomas Cook, the pioneer of package travel, launched the first 'Circular Tour' around the world. The three-month tour packaged the delights of the world – from Europe to the Middle East to Asia and the Americas – as a self-contained tourist destination. Today, the world continues to figure as a destination for round-the-world travellers who coordinate a series of tourist places into an itinerary that 'counts' as global. As round-the-world travellers 'visit planet earth' through corporeal travel, they are increasingly able to translate their experiences of the world into a virtual destination online. With the integration of the internet, laptop computers, and digital cameras into travel practices, round-the-world travellers are now able to create 'real-time' (or more strictly, regularly updated) websites where their friends and families back home are invited to 'play along' in their global adventures. The global becomes a place to play through this interplay between embodied global travel and virtual representations of these experiences.

Drawing on the stories that round-the-world travellers publish on their websites and recount in interviews, this chapter considers the global places and forms of play that the internet has made possible.[1] Websites not only constitute a new kind of tourist destination where people can play online, but they also have implications for how we think about the meaning of corporeal travel. As new mobile communications technologies such as the internet rework the distinction between home and away, work and leisure, and real and virtual, the definition of travel as an escape from the routines and social obligations of the everyday is called into question. Thus websites become a place where travellers also play in between the lines that hold certain categories such as work and leisure, or the real and the virtual, in opposition.

Central to this discussion is the relationship between place and processes of globalization (see chapter 6 above). With metaphors of 'placelessness' (Relph 1976) and 'non-places' (Augé 1995), theorists argue that certain places – such as airports, highways and shopping malls – are emptied of local specificity and rearticulated within global capitalist modernity as homogenized spaces of mobility and consumption. The introduction of the internet has reanimated anxieties over the impact of globalization on our sense of place, with some theorists arguing that virtual space threatens to collapse distinctions between near and far and to overtake real, lived spaces with homogenized, abstracted environments (Nunes 1997; Ostwald 2000; Holmes 2001).

Instead of thinking about global places as non-places or as sites of spatial collapse, this chapter argues that global places can be thought of as those where meanings and boundaries are *in play*. The first section of the chapter describes how the global becomes a place to play in the virtual realm of cyberspace. In particular, it considers the round-the-world website as a virtual place *to* play and as a place *in* play. Following this theme of play and virtuality, the second section goes on to consider how round-the-world websites constitute a stage upon which categorical distinctions are also put into play. This section asks how embodied and virtual travellers are making and remaking the distinctions between categories such as work/leisure and real/virtual, and examines the implications of this boundary play for travellers' mobile sociality.

Playing online

The global has historically been packaged as a place to play not only in round-the-world itineraries, but also in various forms of virtual travel that, in effect, bring the world to the traveller. From the early cinematic devices of the panorama and diorama to the more recent phenomenon of cyberspace, various technologies have been employed to represent and simulate the world itself as a virtual tourist destination. Cyberspace in particular is imagined as a window on the world; a leisure space where travellers can roam the world without ever leaving home. The notion of cybertourism reveals the imagined potential of the internet for virtual mobility and playful activity, with metaphors such as 'surfing' the information 'superhighway' underlining the centrality of travel, mobility, and leisure to our conception of cyberspace. Cybertourism can refer to several types of online practices, from clicking through websites, to taking online virtual tours, to immersing oneself in 'virtual reality' environments.

Before describing round-the-world websites, it is important to situate the consumption and production of these sites in terms of virtual and corporeal travel. Despite predictions that cybertourism may eventually pose a serious challenge to the need and desire for actual travel, researchers have shown that travels in cyperspace are more likely to be used to enhance rather than supplant corporeal travel (Dewailly 1999; Urry 2000). Therefore, it is important to focus not just on travels in cyberspace, but also on the way that virtual travel often involves corporeal travel, and vice versa. Travellers are not only travelling on the internet, but also *with* the internet. Research published in 2002 indicates that the internet and other mobile communications technologies are becoming increasingly integrated into corporeal travel practices (see Wang *et al*. 2002). Producing a round-the-world website is one use travellers find for the internet. While on the road, travellers also log on to investigate their next destination, to send and receive email, to book hotels and transport, to check weather information or currency exchange rates, to participate in online travel forums, and to follow unfolding journeys on other travellers' websites. Round-the-world websites certainly do not provide an exhaustive model of either cybertourism or the incorporation of technology into travel. What they do exemplify, however, is the way corporeal global travel interlinks with and translates into virtual places to play.

Between 2000 and 2003, the major search engines logged over 200 round-the-world websites.[2] For the most part, these websites are personal home pages where independent travellers chronicle the ongoing events of their round-the-world trips. Most websites include general biographical information about the traveller(s) and details about the trip, including maps, itineraries, and information about destinations. The websites also include travel-related information such as travel budgets, packing lists, and health advice. Invariably, these sites also post links to relevant sites, including links to other round-the-world websites.

The bulk of the content of these websites is made up of travellers' journal entries. In these online trip diaries, travellers describe their experiences and illustrate their stories with digital or scanned photographs. These narrative and visual representations of global travel become a virtual terrain where readers at home or at work can daydream about being on safari in Africa, riding the Trans-Siberian Railway, or playing on the beach in Thailand. In this sense the global, as defined by the travellers' round-the-world itinerary and experiences, becomes a virtual place to play for travellers and for their readers.

The regularly posted instalments are intended not only to keep readers updated on the travellers' activities, but also to make readers feel as though they are actually a part of the round-the-world trip. Round-the-world travellers encourage their friends, families, and other 'armchair' travellers to follow their trips via the website and to participate in interactive features such as guest books, online forum discussions, and web polls. Thus round-the-world websites are similar to 'real' tourist destinations in the sense that they are places to hang out, meet other travellers, exchange travel tips and information, swap adventure stories, and see a few 'sights' (in the form of digital photographs).

If round-the-world websites are places *to* play, they are also places *in* play, with fluid meanings and boundaries. I return to the commonly used metaphor of 'surfing the web' to consider how a website might be compared to 'the tube' described by Shields (chapter 5). Like the 'tube' of the surfer's wave, round-the-world websites are mobile, ephemeral, and fluid places. Accessible from almost anywhere in the world, websites appear to be multi-local and mobile sites that move with the traveller. Websites are shifting places, constantly being changed and updated as the traveller journeys around the world. While some websites retain all their content online after the trip is over, their significance as a 'real-time' interactive event for both the travellers and their readers rarely outlives the trip itself. The website's boundaries are also fluid as the hypertext links transport the virtual traveller beyond the site itself (see Mitra and Cohen 1999).

Even this sense of mobility and ephemerality is in play. For many travellers the website actually becomes a marker of stability. The website's URL address may be the only stable address a traveller has throughout the trip. In this sense, the website becomes the traveller's 'home page away from home'; a place to locate the traveller as well as a place of mobility. The website is a place where travellers are always 'present' in some sense, regardless of their movement around the world. Thus, while the website is fluid and mobile, it is also a relatively stable place where connections with friends, family, and other travellers can be maintained.

This kind of 'interactive travel' – with the traveller able to be in touch even while travelling far away – raises questions about the implications of the internet and other mobile communications technologies for the way we travel. What is happening to the distinctions between home and away, or between work and leisure, that have traditionally shaped the social meanings of travel? These questions about the internet's impact on travel fall within a more general enquiry into the relationship between technology and the social world. Recently, theorists have paid much attention to the social repercussions of new technologies such as the internet. Calling into question the status of the body, space, time, and 'the real', such technologies are thought to be reworking the categories around which social life is organized. These technologies are seen as 'facilitating, if not producing, a qualitatively different human experience of dwelling in the world; new articulations of near and far, present and absent, body and technology, self and environment' (Crang *et al.* 1999: 1). Round-the-world websites are one place where new articulations of the categories that have traditionally defined travel and social life are being put into play.

Playing between the lines

Travel has commonly been defined – if not always practised – along a set of fairly clear dualisms. It has traditionally been aligned with leisure as opposed to work, away as opposed to home, and 'escape' as opposed to connection. Contemporary practices of travel – especially interactive travel – no longer fit neatly within these dichotomies. Furthermore, what counts as 'real' travel is thrown into question by the possibility of virtual travel and by the 'virtualization' of tourist spaces (Holmes 2001). While some theorists claim that the relationship between categories is blurring or collapsing, this chapter argues that a more complex coordination between these categories is taking place. In other words, they are in play. In keeping with the themes of leisure and virtuality, I consider two categorical distinctions put into play in round-the-world websites: work/leisure and real/virtual.

The notion that work is now permeated with leisure experiences, and vice versa, is neither new nor unique to the internet. Leisure time has often reflected many of the features of work life (Rojek 1995). However, round-the-world websites provide a useful example for considering how the distinction between work and leisure gets put into play by contemporary travellers. The internet seems to be a place where work and play inevitably merge. In a study of global internet use, Chen *et al.* (2002) discovered that those who are most likely to use the internet for instrumental purposes are also the most likely to use it for recreational purposes. In other words, the more you work on the internet, the more you play on the internet. Thinking about leisure in cyberspace, Bassett and Wilbert (1999: 183) argue that 'the pleasures of new leisure geographies are not an escape from work, but are *connected* to changing work practices', notably the computerization of work.

This assimilation of work and leisure is characteristic of what Wittel (2001) refers to as 'network sociality'. Wittel describes how workspaces have become more playful while leisure spaces have increasingly become places to work. This is especially true for travellers working on updating their websites in leisure spaces like hotel rooms or internet cafés. In the production and consumption of round-the-world websites, the lines between work and play get crossed in various ways. Virtual travellers can pursue the leisure activity of surfing the round-the-world website in their work spaces while round-the-world travellers work during their leisure time to produce these sites.

With website titles such as *Watch Us Wander* or *Backpack With Us*, round-the-world travellers invite the home-bound, or more often office-bound, 'armchair' traveller to become a virtual globe-trotter. Marie, a woman from New York who travelled around the world in 2001, extends the following invitation in the introduction to her website, *Marie's World Tour*:

> Most people don't have the time or money for an extended trip. Now they can travel virtually, visiting every country on my route without a passport....See the world from the comfort of your home or office, for the price of your monthly internet access. Free if you sign on from work.

Marie invites readers to access world travel virtually from home, or even better, from work. From the messages posted to Marie's online guest book, it is clear that many of her readers, such as the one quoted below, take up her invitation:

Seattle

USA

I...can't wait to follow you around the world, albeit from my office desk. You rock!

The 'office desk' becomes a site not only of work, but of pleasurable and leisurely 'virtual travel' as readers follow the traveller's adventures online. In this way, practices categorized as 'leisure' – world travel, surfing the net, and escapism – filter into work-space and -time.

It is worth noting that even though Marie and her on-line 'guest' play with the spatial distinctions between work and leisure, they do not necessarily conflate the two here. For example, the reader from Seattle uses the qualifier 'albeit' to maintain the spatial and conceptual distinction between being at an office desk in Seattle and being on a trip around the world. While Marie and her reader play with the possibility of travelling at your desk, both realize and underscore the difference between taking a real, extended trip that *does* require a passport versus taking a few minutes out of one's work schedule to follow Marie's website. Marie and her reader play with and reassert the distinction between work and leisure.

Just as the internet allows virtual travellers to play in their work-space and -time, access to the internet allows work to infiltrate the leisure-space and -time of a round-the-world trip. Though many travellers quit their jobs or take a sabbatical in order to travel, the decision to maintain a website while travelling reinserts a level of work and responsibility into the traveller's 'time off'. Writing, maintaining, and regularly updating a website while travelling requires an immense amount of work. Websites are the result of 'authentic labour involving the expenditure of effort and energy' (Chandler 1997). Most interactive travellers feel a strong sense of responsibility to keep their websites updated, and that responsibility translates into thinking of the website as a 'job'. Travellers Steve and Leo express their sense of the website as work (interview 1):

Leo: And it's kind of become our job. The website …

Steve: It's the one thing that we have to be really responsible to, that, you know, when we talk about what we have to do it's two things: laundry and the website.

For Steve and Leo, updating the website becomes a logistical chore they have to do to keep the playful project of virtual round-the-world travel going.

Like many round-the-world travellers who maintain websites, Steve and Leo had their website address printed on business cards to hand out to people they meet while travelling. Handing out business cards and scheduling time to update the website brings the objects, rituals, and duties of work into the leisure activity of round-the-world travel. Furthermore, many travellers consider their round-the-world trip and website as an investment in their working future. These travellers plan to include their travel experiences and website production on their work résumés and some even provide links to these on their round-the-world websites for the benefit of potential employers. In this sense, round-the-world travel is seen less as a leisure pursuit separate from the world of work and more as a way of accruing 'travel capital' or 'global social capital' that will enhance the traveller's future career possibilities (see Desforges 1997; Elsrud 1998).

Work can also be a source of pleasure. Travellers find pleasure in doing their websites – a pleasure not least of all derived from the ability to stay connected with their friends and families. In other words, playing with the distinction between work and leisure has a significant social aspect to it. A traveller's conception of his or her website as work is often interwoven with their sense of social obligation to their readers. In fact, the sense of responsibility that Steve and Leo feel towards their website is in some part generated by the demands placed on them by their own social network. Steve and Leo recall the feedback they received from their friends at home (interview 1):

Steve: We started getting all this great commentary back from our friends. 'This is so cool, we're living vicariously through you, it feels like we're getting a little taste of wherever you're going.'

Leo: Then we started getting hate mail.

Steve: Yeah! If we didn't update it like every week.

Leo: 'Where are you guys?' 'What's wrong?' 'I come to work every morning, I check your website.'…

Steve: 'It's been three days since you've updated the site.'

Leo: They want to know what we're doing [and] where we are.

In committing to produce the website, Steve and Leo involve themselves in a social network they feel obliged to maintain along with the website.

While travellers make links between travel and work they also, in several instances, reinforce a clear distinction between nine-to-five office work and the work of travelling and producing a website. Work and leisure are coordinated, but do not necessarily blur into one another. Instead, the boundary between work and leisure is opened up and becomes a place where travellers can make social connections and maintain social obligations. In other words, working during leisure time and playing in work time allows travellers and their readers to stay connected without necessarily confusing their roles and spaces.

As with work and leisure, the internet is seen as threatening the categorical distinction between the real and the virtual. Theorists examining the relation between the virtual and the real have tended to take one of three approaches. First, the earlier accounts of cyberspace envisioned the virtual as a place completely distinct from the real, where users could participate in virtual communities, play with fictional identities, and indeed conduct a separate 'life on the screen' (Turkle 1995; see Rheingold 1994). Second are those theorists who argue that the virtual overcomes physical space and distance. Theorists taking this approach identify the way virtual worlds and spaces annihilate physical space through the speed and reach of information and communications technologies (Nunes 1997; Stratton 2000; Cairncross 2001). The virtualization of tourist spaces is particularly evident in settings such as shopping malls, theme parks, and heritage sites, where homogenized, abstracted, and commodified global spaces replace or displace the local (Rojek 1998; Holmes 2001). This perspective suggests that the distinction between the real and the virtual has become untenable.

In contrast to the notion that the virtual world is separate from the real, or that it replaces the real world, is a third perspective. Many theorists now see the virtual as intricately bound up with the spaces, routines, and relations of the real world (Shields 1996; Bassett and Wilbert 1999; Wellman and Haythornthwaite 2002). As Dodge and Kitchin (2001) insist, 'cyberspace and geographic space are not separate realities, they are interwoven' (24). In other words, the boundary between the real and the virtual is neither impenetrable nor collapsed, but rather 'in play', criss-crossed by a series of practices, mobilities, metaphors, and meanings.

Travellers make and remake the boundaries between the real and the virtual in their websites in ways that allow them to stay interactively connected even while they are on the move. The personal effects of an individual traveller's adventure are transformed into a public and collective experience when travellers post virtual photo albums, 'electronic journals', and 'virtual scrapbooks' of their journeys (interview 1). Also, by employing interactive features such as guest books, interactive forums, web polls, and live online chats on their websites, travellers are able to negotiate between the real and the virtual in ways that involve the audience in the travel experience.

An interesting example of this negotiation between the real and the virtual is *Let me stay for a day*, a website published by Ramon, a student from the Netherlands (see figure 15.1). Ramon planned to hitchhike around the world, but he had no funds to pay for places to stay. To solve this problem, Ramon published a website where people could submit invitations to host him for one day. Through the website, Ramon received thousands of invitations from people in 72 countries and in May 2001, he set off to visit his hosts. Ramon's travels began in Europe and moved on to South Africa, Australia, Hong Kong, and Canada. After spending a day with each host, Ramon would post an entry on his website that included stories and photographs of his host

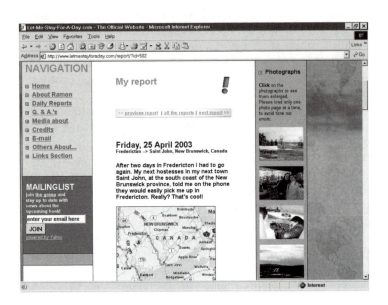

Figure 15.1   One of the daily reports on Ramon's *Let me stay for a day* website

and their activities together. Ramon finally brought his travels to an end and returned home in July 2003.

The real and the virtual are translated back and forth in this process. Ramon's hosts are usually readers who have been travelling virtually through his website. Once they post an invitation and host Ramon for a night, these virtual travellers become participants in Ramon's real travels. In turn, his real experiences with the host become content for the website as the stories and photographs of his visits become the virtual terrain that other virtual travellers explore via the internet. Furthermore, many of Ramon's hosts continue to travel virtually via his website after they have already hosted him.

This movement between real and virtual involves shifting roles of host and guest as well. Ramon is a host on his own website; he welcomes readers to his website and invites them to join him on his travels. However, in his physical travels, he is the guest and is hosted by the same people he hosts as guests on his own website. This circular process of hospitality is actually about generating and maintaining a set of social relations. The website 'introduces' Ramon to the hundreds of people he meets in person during his trip. Through the website, Ramon is in contact with future hosts and is able to stay in touch with past hosts. Furthermore, the forum on Ramon's website allows his hosts and all of his other readers to interact online with Ramon and with each other. His real trip around the world thus becomes something of a virtual travelling community through his website.

Travellers are often very creative in the way they use internet technology to create a 'real' sense of a 'virtual' trip for their readers. For example, on *Marie's World Tour*, Marie offered her visitors actual souvenirs of their virtual tour through her site. In return for a $25 donation deposited into Marie's online account, she would send the reader a real souvenir collected during her own trip. Depending on where Marie was when the order came in, her virtual visitors might receive a shadow puppet from Indonesia or a set of carved wooden monkeys from Thailand. These material objects marked and materialized a journey that took place in between the real and the virtual, simultaneously commemorating a real trip and a virtual trip.

In these examples, the traveller's experiences translate into virtual experiences for the reader. In addition to allowing readers to follow along from home or work, the interactive nature of the website means that the reader can also influence the traveller's experiences. Advice and suggestions posted on forums or registered in 'web polls' often influence the traveller's itinerary. Knowing that the audience is following along compels travellers to do things they might not otherwise do, as Hilary describes (interview 2):

> I think [doing the website] also kind of encouraged us to do things that we might not have done. For example, the balloon trip that we did while we were in Turkey – that was really expensive, but in the backs of our minds we were thinking, and it became a joke, would our readers want us to do this? Yes, they would. So a lot of the things that we did I think were influenced by the fact that we're not just travelling for ourselves, we're doing it for the readers out there.

The things Hilary and her husband did on their trip were influenced by a desire to produce 'good copy' for the audience. They were constantly aware of the fact that their physical experiences would be translated into a virtual global experience for their readers.

At the same time, they appreciated their social obligation to travel – and travel well – for the audience they were 'bringing along' with them, a sentiment echoed by Steve (interview 1):

> It sounds kind of cheesy, but we feel like we're travelling for more than just the two of us. [...] We know that whatever picture we take, it's not going to be just the two of us seeing. There's going to be a couple hundred people seeing it.

Mediating the trip through the website transforms it from an individual experience to a collective experience that involves the audience *during* the trip.

Crang's (1999) analysis of the tourist practice of taking photographs is instructive for thinking about websites as sites of sociality. Photographs do not merely capture a landscape or a moment, but rather are used as a 'communicative tool, to reach people distant in space and time' (245). Much like the tourist photograph, websites mediate a social relation with an absent audience. But whereas the photos in Crang's account are intended for a future audience, the website is aimed at the audience right now.

Travellers share their travel stories almost as they happen, and this timeliness allows the audience to become a part of those stories.

However, this interactive participation must be kept in perspective. Travellers and their readers play with the distinction between the real and the virtual, but this is not to say that this distinction collapses. Both the reader and the traveller acknowledge and reassert the distinction between real travel and virtual travel. For example, participants in one of the 'live chats' hosted on *Marie's World Tour* indicated that although they referred to their visits to Marie's site as 'virtual travel', they did not mistake their experiences there for a real travel experience. In response to my question about how they understood their participation in Marie's site as virtual travel, the online chatters responded with the following comments:

> 'I don't feel like I've been to these places but I do know a lot more about them than I did before.'

> 'I don't think it feels like travelling, but reading Marie's journal makes the locations seem more…down to earth or something. More accessible.'

> 'I haven't felt like I was travelling, but I have felt somehow more connected to these places that were merely names on a map b4.'

These readers felt included in Marie's journey through learning about, accessing, and connecting to places cognitively rather than through the physical sensation of movement through them.

As with work and leisure, the real and the virtual are not blurred in these instances, but travellers and their readers do play with the possibilities opened up by the internet for the audience to be interactively involved in the trip. The making and remaking of the distinctions between work and leisure and between the real and the virtual becomes a site of social interaction.

Conclusion

Round-the-world websites are produced as global places to play through multiple physical and virtual mobilities and by a series of movements across conceptual boundaries. The global becomes a place to play through the interplay between embodied world travel and virtual representations of the world as a tourist destination. At the same time, the possibility of staying in touch through the round-the-world website puts various social and spatial categories into play. Round-the-world websites are places to play, places in play, and places where travellers play with categorical distinctions. What, then, are the implications of this boundary play?

As the extracts from travellers' websites and interviews suggest, travellers play with these distinctions in order to create and maintain their social networks. The boundaries in question are productive places where travellers forge a sense of belonging and connection. For example, work and leisure are interwoven as a way for travellers and their readers to play together. The coordination of work and leisure becomes a site of social connection and social obligation. Similarly, the distinction between real and virtual is reworked in order to make the individual travel experiences more collective.

Through a variety of interactive features, audiences are able to join the round-the-world traveller in his or her global adventures.

Round-the-world websites are constituted by, and make possible, new kinds of interactive social connections, both in person and at a distance. Round-the-world travellers derive a kind of mobile sociality through a combination of movement and connectivity, making them 'nomads...who are always in touch' (Benedikt 1991: 10). Though travellers have always stayed in touch by letter and postcard, the internet allows them to maintain timely connections in ways that were impossible even in the early 1990s, before the worldwide web, the wide availability of laptop computers, and the prevalence of internet cafés. By using the internet in this way, interactive round-the-world travellers are not travelling in order to escape social obligations, but rather travelling as a way of making and maintaining multiple social connections, both in person and at a distance.

This discussion of round-the-world websites points to some of the more general characteristics of global places to play. To begin with, this analysis provides a critique of the way globalization has been imagined to affect our sense of place. Unlike 'non-places' and 'placeless' places that are emptied of meaning or authentic social exchange by the depersonalized mobility of supermodernity (see Augé 1995 and Relph 1976), round-the-world websites are places where people carve out moments of connection and sociality within mobility. Global places to play are places where people seek out social connections of various kinds, whether genealogical (chapter 8), spiritual (chapter 7) or professional (chapter 6). These places are about forging a sense of community and belonging precisely through the mobility – whether online or on-the-road – that characterizes contemporary social life (see Urry 2000, 2003b). As Holmes notes: 'To commute or communicate – both terms being etymological cousins of *communis* [Latin, 'shared, common, universal'] – is to realize community by some form of exchange' (2001: 5). In other words, a sense of belonging can be derived not just by being rooted in a place, but also by moving and communicating between places. As global places to play are produced out of the interconnected mobilities of people, objects, images, and places, they are also producing ways for people, objects, images, and places to be globally connected and connected to the global on both physical and virtual grounds.

List of cited websites and interviews

Cited websites

Backpack With Us (March 2001 to July 2002)
www.backpackwithus.com (accessed 20 June 2002)

Let me stay for a day (May 2001 to July 2003)
www.letmestayforaday.com (accessed 24 April 2002)

Marie's World Tour (January to December 2001)
www.mariesworldtour.com (accessed 15 February 2002; live chat hosted and accessed 17 September 2001)

Watch Us Wander (January to December 2001)
www.watchuswander.com (accessed 27 June 2002)

Cited interviews (all names changed)

Interview 1: 30 July 2001, Steve and Leo, interviewed in person in Berlin during their trip.
Interview 2: 12 July 2002, Hilary, interviewed by phone after her trip.

Notes

1   This chapter draws on a larger research project on round-the-world travel, technology, and globalization. The data is based on interviews with travellers, online participant observation, and in-depth analysis of a sample of 40 round-the-world websites. These websites represent 80 travellers, 45 male and 35 female. The travellers range in age from 7 to 60 at the time of the trip, though most are in their early to mid-thirties. The majority of travellers are from the United States, Australia, Canada, and the United Kingdom, but the sample also includes travellers from countries such as Poland, Norway, Sweden, Germany, and Japan.
2   Round-the-world websites are classified as a subgenre of online travelogues. There are close to 2,000 online travelogues catalogued by the major English-language search engines such as Google, Yahoo! and Altavista.

Chapter 16

'Let's build a palm island!': playfulness in complex times[1]

Mattias Junemo

Introduction: what is this place?

In line with a consistent strategy to develop Dubai's tourism industry, and generally to
enhance the city's image and reputation, the government of Dubai has begun the
construction of the world's two largest man-made islands. They are both shaped like
palm trees and are called the Palm Jumeirah and the Palm Jebel Ali respectively. Even
though they are separate islands, the whole project is referred to as 'the Palm'.

The two islands are located about 24 kilometres from each other along the Dubai
shoreline, about 25 kilometres from the centre of the city. They will be so big that
they can be seen from space: the Palm Jumeirah is 6 kilometres in its longest dimension, and the Palm Jebel Ali will be even bigger, at 7.5 kilometres. Each has three
main sections: the trunk, the crown area with 17 fronds, and a crescent that is
stretching around the trunk and the crown like a barrier to give protection from the
surrounding sea. They are being built from 100 million cubic metres of rock and sand.
The rock is taken from sites within the country, and the sand is dredged from the
seabed. The completion date of the Palm Jumeirah is late 2005 (see Figure 16.1), and
the Palm Jebel Ali should be finished about 18 months later.

Even though the islands are similar in shape they are conceptually different. The
Palm Jebel Ali will have a stronger profile as a tourist resort because it is more activity-
based. The trunk of the Palm Jebel Ali will have a big marine theme park. Because it is
bigger than the Palm Jumeirah, a boardwalk can be fitted in the space between the
crescent and the crown area. The boardwalk spells out a poem in Arabic written by
Sheikh Mohammed bin Rashid Al Maktoum, the initiator of the project. When the
construction is ready about 1,000 homes will be built on stilts along the boardwalk.

The Palm Jumeirah, where land has now been fully reclaimed from the sea, will
largely be residential. The trunk will have apartments, garden homes, and a few high-
rise hotels. At the centre of the crown town houses are to be built, and out on the fronds
signature villas will be placed in more secluded, private neighbourhoods. In total there
will be approximately 2,000 homes. On the crescent surrounding the island there will be
40 plots for hotel developments of about 200 rooms each. Eventually the two islands
together are planned to support an astonishing 100 luxury hotels, 5,000 residential
villas, 3,600 apartments, 4 marinas, the water theme park, restaurants, shopping malls,
sports facilities, themed diving locations, and various other leisure facilities.

This huge project will substantially increase the resources of the tourism industry.
In total it is said that the Palm project will provide Dubai with 120 kilometres of
new beaches – a 166 per cent increase to the existing Dubai coastline. The additional

Figure 16.1   The Palm Jumeirah, Dubai

hotel capacity is being built to meet the dramatic increase in the number of tourists coming to Dubai that is expected. In 2000 Dubai had 2.4 million guests; the prognosis is that by 2010, some 15 million tourists will arrive. If these projections are correct, the Palm will become a huge resource for tourism mobilities (Dubai Economic Report 2003).

A project of this kind can also contribute to other economic sectors, however. In the processes of globalization, places are to a large extent sold as images (Shields 1991). To construct buildings and to restore whole areas, a strong aestheticization is deployed in the fierce competition between destinations (Hall 2000, and see chapter 11) and also to compete for more inward investment (Harvey 1989; Meethan 2001). As a high-profile project the Palm is already being used in Dubai's overall strategic marketing. An already completed and successful flagship is the Burj Al Arab hotel. The distinctive shape of the building, in combination with a promotional narrative of being the world's only 'seven star hotel', has made it one of the most widely known images of Dubai.

The Palm should therefore be understood as part of what globalization entails, both in itself but also as an element of the urban context of Dubai. It will be argued that the Palm can be seen as a case where an aestheticization has been adopted in order to adjust to the interconnected and extensively mobile features of globalization. Essentially the Palm is made visually attractive from a mediatized perspective. In reality it will not be possible to appreciate the extraordinary shape of the islands other than from the air or from space. They are simply too vast and stretch too far out into the sea to be encompassed from a single viewpoint on the ground. The islands are not meant to be gazed at from a geographically proximate vantage point. Instead, as replicated images of an aerial viewpoint, they can be visualized and seen from the other side of the world, or any other place, thus intermixing the real and the virtual (see chapter 15 above). Even though the islands are real places, this reality does not have primacy over their virtual existence. In the Palm sales office they articulated this by saying, 'I think the concept is the most important thing.' Thus what makes the Palm interesting and special as a landmark is the relationship between the reality of the infrastructure and the life worlds premised on mobility and on the ability to transcend time and space.

Consequently it will be argued that the Palm is a landmark and a place corresponding to a society marked by relational and networked connections. The Palm is a place adapted for these circumstances. It can be described within a topology of interconnected spaces, rather than a Euclidean topology of geographically fixed places. From this point of view the Palm is a 'landing mark', marking a spot where flows can land, meet, and connect.

The focus in the chapter is on the aesthetic dimension in relation to what makes mobile life at a distance possible (Urry 2000). Theorists often discuss the material transformations and then focus on the technological aspects of what constitutes global connectivity: telecommunication, fibre-optic cables, satellites, transport systems, credit cards, and so on. Of course the technical infrastructure is essential. As Harvey argues: 'Spatial organization is necessary to overcome space' (Harvey 1985: 145). By this he means that the compression of time and space is necessarily based on the spatial fixity of immobile systems, such as transport, communications, and institutions of law, education, and welfare. But apart from such technological approaches, and the systematic transmissions of flows discussed within these discourses, there are other aspects to consider that are exemplified by the development of the Palm.

Dubai, a place of and for mobility

In the wake of globalization, Dubai has been moulded to become a place adapted for conditions of high mobility and an extensively interconnected world. In a way, the city is an answer to the question: what would a society look like that was developed in the information age, with virtually all the funds needed to realize any vision?

Traditional Dubai's economy was based on fishing, pearling, camel breeding, and regional trade. In 1966 oil resources began to be exploited. At this point Dubai was still part of Trucial Oman, which was under British colonial influence. Not until 1971 did Dubai become one of the emirates of the independent federal nation of the United Arab Emirates (UAE). The country's oil resources took Dubai into a tremendous spiral of development: paved roads, telephone, water supply, and electricity. The massive restructuring of Dubai has been undertaken because its oil resources were comparatively small. This has provided the incentive to diversify its economy in order to become less dependent on oil. Thus the strategy has been systematically to reinvest the revenues from oil into society.

Today Dubai is truly a place of flows. Dubai airport is one of the world's fastest growing. In 2003 about 18 million passengers passed through it, which was an increase of 13 per cent from the year before (itself an increase of 18 per cent on the year before that). The prognosis is that by 2010 the number of passengers will have increased to 40 million (*www.dubaiairport.com*). Another important example is the political initiative to designate certain areas as so-called Free Zones. These are territories considered to be 'offshore' which means that companies here can maintain 100 per cent ownership of their businesses (those registered in Dubai must be at least 51 per cent owned by UAE nationals), are exempt from taxes, there are no customs, and repatriation of capital and profits is unrestricted (www.jafza.co.ae). The purpose of these areas is to boost the economy by minimizing the friction for companies to relocate to Dubai. They also offer adequate infrastructure and resources needed for different types of business.

Consequently, the strategic planning of Dubai since the 1970s has made the city into an economic and cultural hub in the Middle East. The city has become a place where global flows of capital, people, culture, and information land and intersect. Oil now forms only about 10 per cent of the emirate's GDP. The main sectors are finance, trade, and tourism – with tourism contributing about 12 per cent to the total GDP (Dubai Economic Report 2003). The style of leadership behind these achievements indicates a recognition that Dubai is deeply embedded in the flows of the global economy, for instead of seeing globalization as a threat, the society and economy have adapted to these circumstances. For example, in 1999 when Sheikh Mohammed bin Rashid Al Maktoum announced the plans to build Internet City (a free zone for internet-based companies), he said: 'In the future all commercial action will be in cyberspace. But the cyber world will need a ground base on this physical world.'[2]

The way Dubai is being described, it is easy to recognize a clear relationship between the technological infrastructures of the urban space in relation to the conditions of globalization. Castells (1996) is most connected with the notion of a network society. From his perspective the transformation of the urban is directly linked to the emergence of information technology. What he sees being developed are 'informational cities', sites where information infrastructure, educational institutions, and office space become concentrated. This makes cities strategic points of location as the spatial organization is primarily characterized by a logic of the space of flows presiding over the space of places (Castells 1996). As the world becomes more and more integrated, cities are where the flows can be managed and controlled. The previously mentioned Internet City in Dubai is a good example of an informational city. Through offering the very best informational infrastructure, within a free zone, some of the most influential computer companies have located their regional offices there (see *www.dubaiinternetcity.com*).

Another technological aspect is the increased integration of machines of mobility *and* social life. Relations of work, private life, and group belonging are maintained in ways that require different forms of transport and communication. Sociality is occurring at a distance where travel becomes necessary in order to meet others and sustain relationships. Thus, at its core, mobility theory is also focusing on the relationship between mobility and the immobile infrastructure that enables this mobility to occur. In Urry's words, for example, 'the so-far most powerful mobile machine, the aeroplane, requires the largest and most extensive immobility, of the airport city' (2003a: 125). Mobile life is consequently not possible to divorce from the urban infrastructure. Dubai, and its ever-expanding airport, is a good example of this. The mobility that the airport facilitates is also evident. Perhaps the most significant example is that almost 90 per cent of the population are temporary guest workers, constantly arriving and departing in order to sustain their overseas relationships.

These points show the centrality of technology, and how Dubai is embracing these conditions of mobility and interconnectivity. Dubai is made into a place premised on the flows passing through it. It is a trans-local city where resources for development are not just to be found within Dubai (since even oil has lost much of its importance), but come from an ability to attract them (such as workers) from the outside. Rather than trying to secure a border and defend the inside from a potentially threatening exterior environment, its view of the function of the state is as a facilitator and regulator of flows, like a 'gamekeeper', by providing the necessary infrastructure (Urry 2000).

As technology gets incorporated into many aspects of social life, is it then appropriate to characterize life in Dubai using Lash's concept of 'a technological form of life'? Lash's claim is that our own organic systems operate as interfaces towards a technological system (Lash 2001). Or as Amin and Thrift claim, our bodies are engaged with the urban 'mechanosphere' (2002). I would agree with Lash and the other accounts, but would add that the aesthetic dimension is missing. As the Palm project indicates, the aesthetic development in Dubai is just as noteworthy as the technological and the economic. Across Dubai the urban space is being made aesthetically appealing. The façades have spectacular designs and whole areas are thematized. The Palm is only one example: other projects are already completed, such as the Burj al Arab, and more are on the way, including a pre-designed Chinatown; the World (another complex of man-made islands in the shape of a world map); and the Burj Dubai, which is planned to be the world's tallest skyscraper.

The urban landscape is also framed by thematized events such as the month-long Dubai Shopping Festival, and sports events such as the international golf tournament, the Dubai Desert Classic. In 2003 Dubai also hosted the World Bank annual meeting. Thus the technological restructuring is not necessarily leading to a dystopian urban techno-condition of bare, rational, and inhuman systems, as envisioned in Ridley Scott's *Blade Runner* (1982). Therefore Lash's argument of a technological form of life should be extended to note how this mode of living also involves the formation of aesthetic notions, an aesthetic reflexivity (see Lash and Urry 1994).

Technological evolution can be seen as facilitated by the development of aesthetics and design. For example, graphic interfaces have enabled computer technology to be used by a broad mass of the population without special training, as opposed to the expertise in programming languages necessary to use text-based operating systems such as DOS. This is enhanced because graphic interfaces link certain 'physics' to functions for manipulating the software. For example, icons on the screen can be dragged, pulled, and released in a given way defined by the logic of this invented physics. Through such graphic representations of the will of the body it is possible to manipulate the computer system within this created virtual reality. Thus, what distinguishes this from other ways of interacting with previous technologies is that the graphic interface simulates interactions in which the technology is not only manipulated within the physical world, but also within a parallel virtual one. The graphic interface thus produces a virtual representation of the body within the technological system. This, together with how informational systems are organized in networks, makes the graphic interface a 'window' where the body comes into contact with other actor-networks within the space of flows.

The 'aestheticizing' of Dubai is not an exact parallel to the development of graphic interfaces for computers, but urban aesthetics do facilitate connections, and enable sociality in ways similar to the graphic interface. The scope of urban analysis must go beyond the technological dimension and enter the environments created to facilitate this connectivity, to consider how these are provided with an aesthetic coating that provides possibilities and regulations for networks to join together.

Playscapes

The problem with the accounts of the theories focusing on technological transformation is not that they reject or overlook aesthetic transformation. On the contrary, changing urban aesthetics are studied with great interest. Urry explains how changing forms of mobility have made vision the most important of the senses, turning cities and countrysides into 'sceneries' (2000). The impact of tourism has made cities much keener to portray their culture and heritage (Meethan 2001, see chapters 6, 11, and 12). But what these accounts show is that the aesthetic dimension is seen as the outcome of information and mobility technology. It is treated as a dependent variable to more fundamental processes of technological restructuring. But the analogous discussion of graphic interfaces shows that the aesthetic dimension is a completely *integrated* element of the technology. The aesthetic is directly linked to the functioning of the system. In a similar fashion it therefore makes sense to consider the aesthetic environments of cities, not as a consequence, but as part of the very infrastructure that enables mobility and connectivity. There is an economy of signs within technological spaces (Lash and Urry 1994).

In the same way that the graphic interface involves at least a virtual interaction between user and machine, the thematized environments of Dubai provide an interactive milieu. This interactiveness is here considered in relation to what Asplund calls social responsiveness (1987). He explains this term as the basic will to respond when being addressed. He argues that humans have a basic tendency to respond to stimuli. The response is not given like a reflex, however, but is always to some degree ambivalent and uncertain. Social responsiveness is more related to being curious and unprejudiced in finding out about the world. The concrete form that this social responsiveness can take is playing, or being playful.

Ideal playfulness is characterized by being informal, as it does not follow established rules. If there are rules, these are easily abandoned. Playing in its purest sense thus lacks a pre-defined structure and is an open-ended activity of people making immediate improvised responses to the stimuli of environments and of others. Playing also involves a will to participate, which means that responses are immediate and not delayed. Normally, playing involves at least two people, but it can also involve objects if these objects can in any way 'respond'. Within Asplund's theory, being socially responsive in this way is understood as an elementary form of social life. This means that it precedes the consolidation of a group and its behaviour. It is pre-cultural, but is the building block for founding a social group or culture.

The thematized environments of Dubai thus promote a social responsiveness because they are prompting responses to their features. They aggregate this responsivity, not least because their creation is characterized by playful free associations and unrestrained fantasy. To create a place like the Palm is almost an example of divine play: the possibility of creating land at will in any shape. This imaginative play continues as each buyer of a signature villa or beach house has the choice of different architectural themes, such as Contemporary, Arabic, Mediterranean, Caribbean, or Scandinavian (there are 22 styles in all).

This aesthetic of borrowing from around the world and playful juxtaposition seems a general feature of places to play (in chapter 2, Sheller also mentions the use of different 'Eastern' building styles in Caribbean resorts). Similarly, the crescent of the Palm Jumeirah will be a tourist resort designated for 40 hotel lots, each of which will have its own national theme. At the Palm office they said that they 'like to think of it as going around the world without leaving Dubai'. Perhaps this can be seen as an urban materialization of virtual round-the-world travel (chapter 15). The Palm will provide the round-the-world experience without people doing the journey.

This fragmentation of the aesthetic expression is widely recognized within post-modern theory and the playful relation to symbols that lies behind it: symbols apparently can come to signify anything, as the sign has lost any significant relation-ship to the signified (according to Harvey 1989). Baudrillard argues that society has become a symbolic simulacrum of signs only referring to other signs. According to these postmodernists the world is characterized by the acceptance of ephemerality and fragmentation, where people are overwhelmingly concerned with surface images rather than deep meanings (Harvey 1989). The condition of fleeting media images leads to what Featherstone calls an aestheticization of everyday life, concerned with surface appearances rather than roots (1991).

But while recognizing the playful character of society, postmodernism is burdened by relativism. The existential question in this fragmented fleeting condition is no longer epistemological – to find truthfulness about a singular reality. Instead the ques-tions are ontological: 'What world is this? What is to be done in it? Which of my selves is to do it?' (Harvey 1989: 48). The concern in this chapter is the opposite. Instead of play with symbols leading to a relativization of place, I here suggest it contributes to a new absoluteness of place within a relational topology of space.

As with the graphic interface, the aesthetic of the themed environment is merging with the urban infrastructure. Aesthetics merges with technical infrastructure in order to create certain relational windows or gateways that enable mobility and connectivity by providing spaces where people meet. These gateways are the interfaces in the urban landscape that I term 'playscapes'.

Appadurai (1996) uses the term 'scape' generically to describe global flows that are constitutive of imagined worlds. He identifies five such significant flows: ethnoscapes, mediascapes, technoscapes, financescapes, and ideoscapes. Instead of culture being formulated through territorially based social formations such as the nation-state, these globally dispersed scapes are the basis for forming distanced collective imaginings. The scapes are what make collective meanings possible in conditions of fleeting and transformative globalization.

The notion of playscape is suggested here as an additional scape in Appadurai's framework. It is to be thought of as a structure that runs through the fibres of global urban society and promotes the ability to formulate belongings in a global context of sociality at a distance. Playing and being socially responsive is, as Asplund suggested, a pre-cultural phenomenon (1987). Conditions of high mobility arguably involve making new encounters and facing new situations, and this requires a social responsivity. Established relations also need to be maintained and the relational aesthetics of playscapes act as a catalyst for social processes to occur in a context of temporary meetings.

I now elaborate how the Palm will involve playscapes that make sociality possible and places accessible, but also how it creates new boundaries and restrictions upon social life. It depicts ways to find a common ground for collective imaginings when social life is premised on mobilities (and the necessary maintenance of social bonds over great distances) and places are multicultural.

The opening of relational spaces and network sociality

Playscapes create openness and accessibility. We have already mentioned the mobile population of guest workers that stay for shorter or longer periods in Dubai, and who make up some 90 per cent of its total populace. The state must be able to receive these inward flows of new people and deal with a population in constant flux.

In order to accommodate this cosmopolitan mobile population, aestheticization is made neutral by the borrowing or producing of images wherein no single position is privileged. The Palm can be seen as an example of a generic place whose symbolism does not relate to a specific culture. One interpretation is as good as another in the purely thematized environments on the islands. Thus, as one Swedish family told me, Dubai is an easy place to adapt to. They thought it would have been more difficult if they had instead moved to Germany, where they would have faced a much more established German culture. In Dubai they are instead facing fleeting and undefined cultural expression.

This sense of openness is created through the fact that the symbolisms at play have lost their moorings in locality. What this means is that the places are to some degree under-defined, or within Asplund's terminology they are elementary pre-cultural places. Shields (1991) describes this feature in his discussion of West Edmonton Mall, where Paris is used as a thematic model. By bringing in the symbolism of Paris, the bond is broken between the local geographical positioning (the city of Edmonton in Alberta, Canada) and the performances taking place. People engage in certain forms of shopping and *flânerie* that relate to understandings of what Paris is like. This knowledge is catalysed through the material surroundings, as people mobilize a performance fitted for the theme. The under-defined thematic environment is unfinished as a cultural consolidation, so that its scenes can be completed through the practices played out there. Coleman and Crang similarly argue that tourist places are performed rather than given: places are part and parcel of the events taking place, not so much the already existing settings for events as the production of space through the event (2002a).

The openness that Dubai depends on is thus created through borrowed or produced (like the Palm) generic images. These do not provide a coherent culture, but do provide openness for people to come and assemble their own meanings. Strolling on the Palm will involve playing with associations of what is already known about the cultural themes, which will be hugely different depending on one's cultural background. The way that Dubai is adapting to the mobility context of an ever-changing population is by making people part of a constant creation, an ongoing play.

The second aspect of playscapes is that apart from providing a cultural openness adapted to mobile conditions, they also become strategic sites for establishing relationships under these mobile conditions. The starting point for the discussion is thus to try to understand the conditions for sociality in a network society. Wittel translates the macro-perspective of Castells' network society into a micro-sociology of the information age (2001). He suggests that there is a shift away from sociality in closed systems and towards regimes of sociality in open social systems – such as networks. In a study of urban media workers he noticed that places of leisure such as pubs, restaurants, golf courses, and especially parties, were strategic places for the practice of 'networking'. He suggests that this entails a shift towards the production of social relations, and away from their *re*production. The use of cafés and pubs for professional meetings allows for a spontaneity that might spark new ideas, and for accidental encounters that might extend networks. The sociality that Wittel depicts between people is more informational and ephemeral, but at the same time it is more intense as it assimilates work and play. Another new element in this sociality is an increasing commodification of social relationships as social capital, signalling a move from having relationships towards making and managing relationships.

Thus playscapes are strategic points to expand one's network, or to manage the already existing ones. It may seem contradictory that just when there have never been greater possibilities to replace physical travel with telecommunications, corporeal presence becomes more important than ever. The important factor of playscapes is that they mediate these meetings, and enable the essential factor of producing and nurturing trust, without which networks would be impossible to sustain. In this way Boden and Molotch argue that memos, fax and email are less effective at establishing long-term trust relations partly because they are more functional and task-oriented (1994; Urry 2002b). Networks cannot be sustained solely by communication, but depend hugely on corporeal presence where the parties have the ability to read off each other's facial expressions and see each other eye-to-eye. As Urry puts it: 'Social life thus involves variously organized "tight social worlds", of rich, thick co-presence, where trust is an accomplishment of such meetings and which facilitates disembedded network sociality' (2003b: 168–69).

The 'thick co-presences' that Urry speaks of can be established through the out-of-the-ordinary-experiences that playscapes provide. After long gaps between meetings it is possible to refer to the collective memories of these shared experiences. The mutual experiences become elements in people's biographies, hooks on which to pin their intersecting trajectories as evidence of their co-presences as they live life *en route*. In this way relationships are pinpointed in time and space in relation to a traceable chain of events – not free-floating, but attached to particular environments. Playful places thus create a shared bond beyond instrumental interests, enabling trust. The notion of playscape highlights that the environment itself becomes mobilized and plays an active part in the creation of social bonding.

For such purposes the Palm is filled with opportunities to meet. Public spaces are very much arranged to provide strong experiences where people can meet and reinforce their relationships. These are crucial sites for establishing a social network, especially for the many Western women in Dubai who are housewives and would otherwise be isolated by the social and legal restrictions placed on women's activities and movements in Dubai. The supply of meeting-places provides informal spaces for bonding, finding opportunities for meaningful work or projects, or information about jobs, schools, events, laws, and regulations.

The third point is that playscapes are often guarded, as network cores try to protect themselves from unwanted elements. Even if playscapes are principally open, they are in reality closed off for many and under heavy surveillance to ensure that this is maintained. Graham and Marvin (2001) point out that places for leisure and consumption are carefully controlled. Shopping malls and theme parks are regulated by closed-circuit TV cameras, private security, and forms of visitor regulation. These places customize their infrastructural connections with the wider urban region. This is shown in the case of the Palm. In Dubai there is a tradition of making places into gated communities, and this will be the case in the residential areas of the Palm. This makes the whole crown area a secluded haven for the people living there. The trunk will be more public, and also the crescent, as this is where hotels and other public amenities will be situated; nevertheless these hotels and apartments will have their own security in the lobby.

Surveillance and control are also maintained through symbolic distinction (Bourdieu 1984) between those with access to the networks and those without. Those inhabiting the centre of a network distinguish themselves from the excluded 'lower classes' through a certain consumer lifestyle. In Dubai this phenomenon even has a name, 'Jumeirah Jane', describing the affluent Western women with a consumer lifestyle represented by mobile phone, sunglasses, a four-wheel-drive car, and colour-coordinated outfits. This can be contrasted with the working clothes of an Indian or Pakistani working in construction or perhaps as a caretaker. Jumeirah Jane is a person who belongs in the Palm, while the maintenance workers are uniformed in order to signal that they are allowed inside, but are not part of the social activities. In other words, apart from simply possessing monetary capital, this must be turned into some accepted form of social capital in order to gain access to the social network.

The playscapes are thus both points of entrance and points of control. They mark the physical entrance point where networks can be accessed, as well as the gate where the socially accepted capital for access must be registered. Even if one can be physically close or actually within the playscape, the inability to enter into its networks is as great as if one were a thousand kilometres away.

Conclusion

The overall theme of this chapter has been to demonstrate the parallel development between aesthetic and technical infrastructures that enable global time-space compression. The Palm has been used as an example of how Dubai is adapting to the conditions set by global mobility and connectivity. The concept used is that of 'playscape', to demonstrate first how a principal openness is created. Second, these spaces are where network relationships are established and maintained. Playful places become strategic places for people to meet and establish trusting relationships. Networks are not purely technical but also human, and in order for human connectivity to occur trust is needed. Connections are not the same as plugging in the power cord of a machine, but rather must be negotiated and formulated in each specific case. Third, at the same time that playscapes facilitate connectivity, they also exclude groups. As Castells (1996) among others shows, globalization is an uneven process. This unevenness is highly visible in Dubai as much as in other cities. Some people have more capital to gain access into the sites of power within the network society. Playscapes are one such dividing line.

This then is a new agenda for tourism research, to think about how tourist sites become part of activities that are not mainly tourism-related. Dubai is a good example of how the line between urban planning and tourism development is more or less indistinguishable.

Notes

1   This chapter is based on interviews with Swedish families made in autumn 1999. I recently returned to interview people at the Palm office, and also other people with insights into tourism development and Dubai society in general. The following are useful internet sources for general information: *www.dubaiairport.com*; *www.dubaiinternetcity.com*; and *www.jafza.co.ae*

2   BBC News. See *http://news.bbc.co.uk/1/hi/world/middle_east/493915.stm*

Chapter 17

Atomica world: the place of nuclear tourism

Kathleen Sullivan

> It was kind of inconceivable as to what we were looking at there. We passed comments back and forth in the plane. We took pictures. By the time we had done that I became concerned that we better quit being sightseers and get out of there. And we were gone and off of the coast in about 20 minutes after the bomb was released.
>
> Paul Tibbets, American Air Force Pilot (quoted in Rafferty 1982)

Introduction

Paul Tibbets, the pilot of the first nuclear sortie in August 1945, refers to himself as a sightseer, thus making the crew of the *Enola Gay* the first, self-proclaimed nuclear tourists. They were viewing from a safe distance the utterly destroyed city of Hiroshima, watching a spectacle that they ignited, which claimed the lives of 100,000 people with a flash of light. Participating from afar, being the watchers safely removed, gazing upon an assortment of events, they defined the characteristics of nuclear tourism. After they took pictures, the men of the *Enola Gay* flew back to base for a beer party held in their honour.[1]

Since Hiroshima, nuclear tourism has been promoted by opponents and proponents of nuclear technology. In this chapter I explore the visible and invisible meanings of nuclear tourism through two contrasting projects: the Sellafield Visitors' Centre in the United Kingdom and the Bravo 20 proposal for a National Park in the US state of Nevada. These are actual and proposed sites that bear witness to the ongoing saga of nuclear technology and the military – unlike the case of Tibbets where being a tourist at the time of a nuclear bomb run can only happen once (and possibly after that, never again). The tourist 'destinations' articulate the difference and interplay between the virtual and the real. Bravo 20 will provide the virtual tour of a bombing range as an educational tool promoting disarmament and an ethic of environmental stewardship: Sellafield Visitors' Centre is a public relations effort to hide the real dangers posed by nuclear energy production. British Nuclear Fuels operates Sellafield, on the Irish Sea coast in Cumbria, north-west England, to reprocess plutonium for use in reactors. However, the plant was initially established to produce fissile materials for the British nuclear arsenal. I begin with a definition of nuclear tourism, and then provide a longer case study of the Sellafield Visitors' Centre. In concluding with Bravo 20, and the story of military activities in Nevada, I point to the merits of a kind of educational tourism.

Defining nuclear tourism

Nuclear museums and sites have long been identified as tourist destinations. Some people visit them as places that symbolize death, while others are fascinated by scientific progress. The Bureau of Atomic Tourism[2] provides a good starting point for locating nuclear tourism destinations. The website was conceived to promote 'locations around the world that have either been the site of atomic explosions, display exhibits on the development of atomic devices, or contain vehicles that were designed to deliver atomic weapons'. The sites listed offer disparate destinations for the nuclear tourist, from the Bradbury Space and Science Museum in the US, which glorifies the development of nuclear weapons as 'peace-keepers', to the Atomic Bomb Memorials of Hiroshima and Nagasaki that provide physical evidence and personal stories of nuclear weapons as instruments of 'omnicide' (Bertell 1985; see figure 17.1).[3]

At the same time, some nuclear tourists may not know that they are nuclear tourists. These unwitting persons frequent radioactive landscapes that are still defined as pleasure zones. This is true for many sites adjacent to nuclear power plants where families come to picnic and play in the natural environment. One such place is the power plant in Tarragona, Spain. This mid-coast beach, located within a stone's throw of two operating reactors, is a popular destination for holidaymakers, complete with a caravan park for summer lets, and other accoutrements of seaside activities.[4]

For this chapter, nuclear tourism is defined as visiting sites where nuclear weapons and power are or have been manufactured. This 'nuclear tourism' is informed by

Figure 17.1   Hiroshima memorial

discourse on tourism provided by Urry who utilizes a Foucaultian model of 'the gaze', which he describes, in part, as:

> consuming goods and services which are in some sense unnecessary...And yet part at least of that experience is to gaze upon or view a set of different scenes, of landscapes or townscapes which are out of the ordinary...[We] gaze at what we encounter.
>
> (1990: 1)

In particular the tourist gaze is 'constructed through signs' (Urry 1990: 3; and see MacCannell 1976).

Signs for the nuclear tourist gaze are often split in opposing directions, 'sparking a reaction' that can be predictable and contradictory. For those who view nuclear technology as manufacturing mass death and contamination, signs of 'death' and 'destruction' permeate their gaze. For the nuclear tourist who views the industry as a symbol of scientific progress, the definitive sign is that of 'safety', which implies other signs, such as 'responsibility' and 'security'.

Within the industry-sponsored visitors' centres dotted across the globe at sites of weapons and power manufacturing, the nuclear tourist gaze is 'not left to chance'; people are instructed on 'how, when and where to "gaze"' (Urry 1990: 9). There are nuclear tourists who will look away when the 'surrogate parents' of nuclear tourism, predominately defined by industry's public relations personnel, attempt to 'protect him/her from harsh reality' and 'restrict the tourist to...certain approved objects of the tourist gaze' (Urry 1990: 7–8).[5] At the Sellafield Visitors' Centre, 'approved objects of the gaze' have been confined to signs of 'safety', and more recently signified by a 'right to know more' and to 'have your say'. This is new to Sellafield, and a radical departure from the previous Disney-like theme park. Through the sign 'safety', the UK nuclear industry is also read as providing information and 'transparency' about the dangers of producing nuclear power.

However, at the competing tourist trap of Bravo 20, 'the consumption of goods and services' as information and signs of safety are turned on their radioactive heads. Urry notes that not all tourists will read the signs provided as authentic. He uses the example of 'post-tourists' who 'find pleasure in the multiplicity of tourist games. They know that there is no authentic tourist experience, that there are merely a series of games or texts that can be played' (Urry 1990: 11). Nuclear tourism provides some fluidity for those knowing and unknowing tourists, as the journey between Sellafield and Bravo 20 will demonstrate.

Touring the Complex: Sellafield as a tourist trap for nuclear misunderstanding

The image of an awestruck young boy greets nuclear tourists from one of many glossy brochures in the foyer of the Sellafield Visitors' Centre. It reads: *Welcome to Sellafield: It's A Great Day Out*. This and many other 'signs' define it as a place for education about energy, a place where parents feel safe enough to bring their children, and where in 2001 over 81,000 visitors stopped for a look around. And the visitors keep on coming, for Sellafield is open almost every day of the year, inviting nuclear tourists with an offer of sheltered accommodation along the windswept Cumbrian coast.

Sellafield is ideally located as a tourist attraction due to its close proximity to the Lake District, a popular holiday destination in the north of England. Tourism is such an established part of the region that within the Sellafield Visitors' Centre there is a Tourist Information Centre that, in 1995, won an award for 'Regional Tourist Information Centre of the Year' (phone interview 28 June 1998). According to a 1997 British Nuclear Fuels' (BNFL) survey, 50 per cent of all visitors cite their reason for coming to Sellafield as being 'on holiday in the area' (BNFL 1997: 4).

There has been an established Visitors' Centre at Sellafield since 1982. The first Centre was relatively small, although it still commanded 13,000 visits a year. In 1988, BNFL spent £5 million to build a new, larger Visitors' Centre; and in 1995 spent another £5 million on the tourist site. According to BNFL's research 'over 400,000 visits were made to a nuclear industry Visitors' Centre in 1992 [in the UK]. Of these, over 145,000 were to Sellafield' (BNFL 1997: 2). In 1993 the Centre had 119,472 visitors.[6]

On 18 February 2003, the British Deputy Prime Minister John Prescott opened the new Visitors' Centre at Sellafield, the 'Sparking Reaction' exhibit. The new exhibition is a far cry from the narrative of the previous exhibition, called 'The Mighty Atom'. This featured a Disneyesque character who used to 'accompany' the visitor from video screens and automated displays throughout the Centre, 'explaining' the nuclear fuel cycle and the development of nuclear physics. Speaking in a regional accent, which suggested its connection to the local community, the Mighty Atom was constructed as local and non-expert, and appeared to be appealing to a younger audience. The Mighty Atom trivialized the dangers of nuclear technology, and offered a confusing narrative about the very real dangers on site, and the initial remit of the Sellafield complex.

In fact, Sellafield's primary function was the production of fissile material for Britain's nuclear arsenal. In 1947, two years after the United States used nuclear weapons against the people of Hiroshima and Nagasaki, work began on the plutonium production piles at the site, then called Windscale (Bolter 1996: 37). Throughout its fifty-year history, Sellafield has hosted a myriad of functions. It is where nuclear weapons materials were produced, along with nuclear fuel, nuclear energy, and its most hailed and failed project, the THORP reprocessing plant, built to 'recycle' plutonium for use in reactors. Now slated for shutdown, British Nuclear Fuels' plutonium fuel production at THORP will be decommissioned out of the public purse, and leave behind a vast legacy of British, Japanese, and German nuclear waste. Sellafield has thus been described by Harold Bolter, former director of BNFL, as 'the world's nuclear dustbin' (1996). How has a nuclear dustbin been turned into a place for visitors?

In its new incarnation as 'Sparking Reaction', the Visitors' Centre provides nuclear tourists with information on nuclear weapons, as well as societal responses to the nuclear arms race, including a brief description of the British peace movement. In one display case entitled Nuclear Weapons, the role of Windscale is cited, but Sellafield itself is not directly implicated in the production of Britain's nuclear arsenal.

> In the 1940s Britain was racing to develop its own nuclear weapons. Reactors *like those at Windscale* (which is now part of the Sellafield site) were built to produce plutonium for nuclear weapons. But soon scientists realized that reactors could also be used to generate electricity, and the nuclear power industry was born.
>
> (Text from display recorded on 8 October 2003, emphasis added)

To merely mention that reactors 'like those at Windscale' were used to produce plutonium does not go far enough in explaining Sellafield's role in the production of British
nuclear bombs. Bolter states that 'Sellafield grew out of the need to produce materials
for the atomic bomb' (1996: 29). This rather more transparent description of
Sellafield's role in nuclear bomb-building is missing from the exhibition. It is not
made entirely clear that the actual reactors at Windscale were designed, built, and
used for the production of fissile materials for the British nuclear arsenal. However,
the discerning tourist will notice that within the single display case on nuclear
weapons is E. P. Thompson's foundational book *Protest and Survive*, which details the
precise role of Sellafield, once called Windscale, in nuclear weapons production.

This current layout and design of the Visitors' Centre, a collaboration with the
Science Museum in London, raises questions and includes controversial information. Is
nuclear power safe? What do we do with nuclear waste? How do we support our
energy needs? What sort of energy should we produce, knowing that Global Warming
is due to fossil-fuel emissions? The 'Sparking Reaction' exhibit provides a more critical
analysis of the nuclear fuel cycle, including controversial issues such as energy production and nuclear waste disposal. This is a major departure from previous exhibitions,
which simply supported the project of nuclear power production.

The aim of the new exhibition is described in the brochure as follows:

> Welcome to sparking reaction, a unique exploration of the debate surrounding
> how we should generate electricity in the UK and the place of nuclear power
> within it.
>
> Use the interactive exhibits to explore the different ways we could generate our
> electricity and the issues surrounding nuclear power. Can we build more wind
> farms? Should we import more gas from abroad? How safe is nuclear power and
> does it still have a role to play? Should we just use less electricity?
>
> Have fun exploring these questions and how they affect your life, decide what
> you think and have your say.

While the exploration of controversial issues is quite unexpected given the role of
Sellafield in nuclear energy production, the idea that learning about energy can be part
of a fun and exciting day out trivializes the dangers inherent to radioactive contamination and the timescapes of nuclear materials. One need look no farther than the hill
farmers of North Wales who, affected by the on-going social and ecological tragedy of
Chernobyl, were unable to sell their sheep in the aftermath of a nuclear power accident
that occurred thousands of kilometres away in Ukraine. Adam notes that 'more than
nine years after the accident, some twenty farms in North Wales were still under quarantine and new "hot spots" were still being discovered' (1998: 199). Adam argues that
we do not have the appropriate theories of time from previous thinkers and theoreti-
cians to assist in analysing invisible, near-timeless poisons such as long-lived radiation,
and this affects how we define it as a 'risk', and ultimately how we respond to the
dangers inherent in its production and use in nuclear reactors.

'Sparking Reaction' does contain a display on the Chernobyl tragedy that states that
'twenty-eight people died from radiation sickness within weeks of the explosion and
scientists predict that 8,000 to 10,000 cancers will eventually be caused by the accident'. This notion of 'eventually' is striking close to a 'temporal critique', but not close

enough. Adam argues that the outcome of Chernobyl 'was, and still is more than ten years later, environmental destruction of enormous spatial and temporal proportions and untold numbers of *dead in degrees*: there and then, a bit later, tomorrow, and some time in some elastic future somewhere' (1998: 198).

Sparking Reaction does not mention the rise in genetic mutations occurring around Chernobyl. Because of Chernobyl, it is estimated that the likelihood of a child being born with genetic mutations has increased by 250 per cent (Roche 1996: 90; Cheney 1995: 190). In a hospital in the Soligorsk region of Belarus, a doctor reported the birth of 22 babies, 'most of whom had been born with no organs' (Roche 1996: 90). Another Belarussian doctor asks: 'Will future generations be "normal", or will they have some kind of serious effect? This is the fear which affects people constantly' (Hoban *et al.* 1994).

Nuclear contamination affects the health and wellbeing of people all over the world, and yet there is no formal role for citizens to influence public policy on contamination, with or without a tourist destination that encourages visitors to register their concerns.[7] What the Science Museum has produced for BNFL relies heavily on interactive components, which suggest that the public is playing a role in decision-making and critical analysis about energy production and nuclear waste disposal. Although encouraged to 'have your say' throughout the exhibit, the idea that registering an opinion at the Visitors' Centre could have an impact on decision-making processes is false.

'Sparking Reaction' offers critical assessments of Sellafield's activities, but the speed with which these messages are displayed blunts their impact. On the ground level of the main space is a scrolling text, illuminated and rolling on the floor where the viewer stands. The tourist becomes a part of the wordscape that is created, as if embodying the whole spectrum of opinion. The words and phrases are variously in support of or denouncing the project of nuclear energy. But they illuminate the contrasting opinions at such a rate, that the reader/participant is left with a feeling of having to keep up. And the words are given barely enough time to be read, let alone penetrate the consciousness of the person standing in the light of the rotating information. It felt more like a game of 'push you–pull me' than critical inquiry.

Speed is also a defining factor of the interactive games dotted throughout the Centre. In one, the object is nuclear waste disposal – or rather, how to quickly bury nuclear materials. In the game, the player is to load lorries with as much material as possible, as quickly as possible, and to get the radioactive materials to a 'safe' repository, deep under ground. Speed is of the essence. And as in real life, there is no effective programme for incorporating viewpoints of the local/affected community. The game is basically about moving material, not what the material is or how exactly to isolate it from the environment.

Dr Rachel Western, a nuclear researcher from Friends of the Earth, London, and a consultant to the UK Radioactive Waste Executive, NIREX, was employed by the Science Museum (and paid by BNFL) to comment on the text of the exhibition. She claimed that the language about radiation is unhelpful for a general audience.

> The text on radiation was too complicated, generally, and the nuclear language around radiation too complex. There is a sort of conveyor belt language that takes the harm of radiation away from an everyday understanding, and I don't think this approach is useful for increasing public awareness.
>
> (phone interview, 4 October 2003)

If information about radiation is therefore being given at a rate at which it cannot be assimilated and in a language that does not communicate its inherent dangers, so what presents itself as a public service or an exercise in transparency is neither. After reviewing the texts for 'Sparking Reaction', and not seeing the necessary changes she had advised, Dr Western refused to put her name to the project.

Although 'Sparking Reaction' is a new feature to the Sellafield experience, one of the most sought-after components of the tourist trap has been lost. Before the terrorist attacks in the United States on 11 September 2001, nuclear tourists were given a bus tour of the site. Now due to security, the 'Sellafield Sightseer', a custom-built coach complete with video monitors that provided short descriptions of buildings highlighted on the tour, has been put out to pasture. This sight-seeing mechanism of the inner workings of a vast industrial complex is a regrettable loss since it highlighted a contrary thesis to that of safety and security.

On the bus tour, the nuclear tourist was presented with an industrial landscape of windowless buildings, labelled by letters and numbers, in an environment signified by security measures to ward off danger or intrusion. Here the gaze was unwittingly directed towards the trefoil radioactive symbol, chain-link fences, guard-posts, and other signs of a militarized industrial estate. The earlier opportunity to be driven into the inner sanctum of Sellafield as a nuclear complex, with its armed security forces and huge technological fortresses, revealed an environment where 'danger' and 'complexity' are the signs to be read.

What the tourist gaze cannot be directed towards are the invisible poisons – radioactive effluent – that for many years Sellafield dumped into the Irish Sea. These deliberate spills are having adverse effects on other nations – the rocky shorelines of Ireland, the fisheries of Norway – whose governments are demanding that the practice cease (*Guardian* 18 July 1998). And because of the uniquely long-lived nature of this waste, far-off lands acquire little radioactive bits of Sellafield that have the ability to cause cancer and ill-health into a distant future. It's like the badges handed out at the Visitors' Centre – 'I've been to Sellafield' – in the form of technetium traces in rock lobsters, and leukaemia clusters in Dundalk.

Bravo 20: the meaning of nuclear war

Meanwhile in Nevada, an altogether different tourist attraction is being proposed. The story of this would-be National Park centres around three men, a dog and their love for a devastated place of wilderness turned into an immense munitions graveyard. Bravo 20, part of an active bombing range of the Fallon Naval Air Station, is a nuclear tourist destination because of its association with the US Government's nuclear stake in the state of Nevada, and its potential role of educating for nuclear disarmament. The Pentagon and the Department of Energy (the US nuclear weapons production agency) have a far reach into the wild desert.

Like the state of New Mexico, Nevada hosts many hallmarks of US atomic history, sure stops on the nuclear tourist trail.

- On the Nevada border with Utah, Wendover Air Force Base is where Paul Tibbets and his crew trained for their mission over Hiroshima.

- The Nevada Test Site, one-third larger than the state of Rhode Island, is the US Government's nuclear proving ground. Over 700 nuclear 'test' explosions were detonated on this land, the tribal home of the Western Shoshone nation. From this single region, the mobilities of radiation can be traced. Radioactivity released from atmospheric nuclear testing – including plutonium, strontium, caesium, carbon-14, and radioactive iodine – have been widely dispersed throughout the world. Underground test explosions have contaminated soil and groundwater. A 1991 US government report called the soil contamination from underground testing at the Nevada Test Site 'a threat to human health and the environment'.

- Surrounding the Test Site is the Nellis Air Force Range, which comprises three million acres of land (12,140 square kilometres) within a larger area of ten million acres (40,500 square kilometres) above which the air space is militarily controlled – the latter figure is, astonishingly, 'roughly the size of Switzerland' (Misrach 1990: 8).

- Also in Nevada, the US Department of Energy has chosen Yucca Mountain, which is on the land of the Paiute nation, as a repository for high-level radioactive waste from US civilian nuclear power plants. At an estimated cost of $58 billion, the US DOE intends to bury 70,000 metric tons of high-level nuclear waste in Yucca Mountain, which is located in an area of massive seismic activity. The mountain has at least 35 active earthquake fault lines running through it. According to the US Congress's Nuclear Waste Technical Review Board (NWTRB), 'many of the DOE's assumptions regarding Yucca Mountain are extreme and unrealistic' (30 January 2002). Still the DOE contends that the high-level radioactive materials buried at Yucca Mountain will remain safe for 10,000 years.

- Finally, at Fallon Naval Air Station, the delivery systems for nuclear weapons, such as Tomahawk Missiles, have been tested. The pilots who carried nuclear loads during the Cold War were trained there. Supersonic engines – with their attendant sonic booms – have been tested over this wide landscape. And these practices are carried out under the national banner of 'Peace is Our Profession'.

In 1986, Richard Misrach, an award-winning photographer, first met one Doctor Bargen. Misrach had hoped to gather information for a project he was working on about military activity in Nevada. He wanted to document, as he put it, 'night raids on rural residents by Navy helicopters, laser-burned cows, the bombing of historic towns and the unbearable super-sonic flights over rural America' (1990: xiii). He might have left his research there, as an observant non-participant, if he had not become intrigued by Bargen's story of an illegal bombing range on public land nearby.

Bargen, a Canadian doctor who shared his love of flying aeroplanes with Dick Holmes, another companion on this initial journey, arranged for Misrach to visit the site called Bravo 20. This first excursion introduced Misrach to the terrible beauty of the bombing range – the awesomeness of the desert, littered with exploded and unexploded munitions, bomb casings, and craters full of blood-red liquid. Describing the view as the distance closed between the three men and Lone Rock – a small mountain bombed to a third of its size by the US military – Misrach describes this place of absence:

As we speed towards Lone Rock, the cars kicked up huge billowing clouds of fine dust. After several miles of nothing we came upon the first bomb. Then a crater.

Then more craters and more bombs. As Lone Rock turned from a bump to a mountain, the playa transformed from pure desert wilderness to the post-apocalyptic landscape of a Mad Max scenario. Soon there was not an acre of land that was not riddled with crater upon crater, shrapnel, and bombs (practice and live). As far as the eye could see in any direction was man-wreaked devastation.

(Misrach 1990: xiii)

The concerns about Bravo 20 for Bargen and Holmes, who met as fellow pilots, had to do with two Notices of Intent published in the local newspaper by the US Navy announcing the withdrawal of 181,000 acres (732 square kilometres) of public land for an expansion of a bombing range, and a proposal for a Supersonic Operations Area (SOA), covering some 14,250 square kilometres, an area about the size of the state of Connecticut. The Navy's expansion and new operations in the area concerned many local people. Bargen was particularly concerned as he often flew to visit his patients in remote areas of the region, and his travel would be restricted and in many cases denied if the Navy had its way. Additionally, the supersonic overflights had known medical and psychological effects, which alarmed the men. In order to address these issues, Bargen and Holmes led a public education effort, which resulted in organizing local people in opposition to the military's objectives. After their exceptional effort over many years before their meeting with Misrach, the Military Land Withdrawal Act of 1986 was passed. This requires the military to conduct environmental assessments and to clean up environmental damage, as well as obliging them to renew their claims on public lands every fifteen years.

To celebrate the passage of the Military Land Withdrawal Act, Holmes, Misrach, and Bargen planted an American flag on the top of Lone Rock. For although the Act had been passed, Bravo 20 had been re-annexed by the Navy and bombing could resume at any time. The men thought that the presence of a national symbol would spare Lone Rock. For Misrach, the view from the bombed peak offered a vision of something more. He saw a National Park there, which would educate people about the military abuse of great expanses of the American West.

Misrach brought together a team of architects to make pictorial renderings of the proposed site, suggesting an appearance of Bravo 20, and which could themselves also be used as a tool for education. This proposal, including the story of the legal battle for public land use in Nevada and other western states, along with Misrach's stunning photographs of the site, are part of the public record through his book, *Bravo 20: The Bombing of the American West* (1990). The site itself has yet to be built, primarily due to a further extension of the US military's use of Bravo 20, which at the time of writing is still an active bombing range.

The proposed tourist attraction would include a Visitors' Center and Museum, a camping area, gift shop, café, walking tour, and driving tour (see figure 17.2). The Visitors' Center will serve to educate the public about the social and ecological hazards of nuclear weapons. Misrach's intention for the museum makes connections beyond Nevada to other regions where the US military has taken land in pursuit of bigger bombs and faster delivery systems.

The Bravo 20 Visitors' Center and Museum will be devoted to the history of military abuse in peacetime. Displays and exhibits will include our radioactive

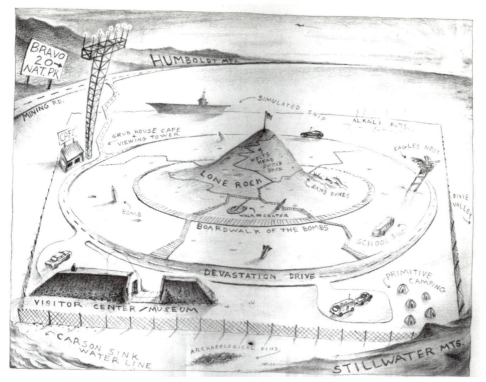

Figure 17.2   Bravo 20 plans

experiments on the residents of the Marshall Islands in the Pacific, the contamination of continental America by tests at the Nevada Nuclear Test Site, the Colorado Rocky Flats nuclear weapons plant and Hanford nuclear area in Washington State, chemical weapons storage, toxic waste disposal, and the confiscation of land and airspace throughout America.

(Misrach 1990: 95)

The Visitors' Center, styled after a military munitions bunker, will include a film and video archive, and a library with material detailing the US Government's atomic and military history. There is also space dedicated for conferences and gatherings of concerned citizens, to use the site as an inspiration to work towards the goal of nuclear abolition and environmental restoration.

With particular relevance to Bravo 20, the museum design will highlight the history of the Navy's illegal bombing activities, and the local community's struggle to reclaim the land. To do this, the physical evidence of the devastation will be literally carved out for a public view.

> The Visitors Center and Museum will have a subterranean level, walled on one side by a huge aquarium-like window. Visitors will be able to view the Lone Rock landscape from the surface level, then descend underground and witness the bombs (replaced with dummies) that have been buried over the years [see figure 17.3]. This feature will call attention to the Navy's flawed environmental studies

Figure 17.3   Visitors at Bravo 20

and arguments before Congress that it would be too costly and too difficult to locate and defuse all the unexploded bombs buried underground.

(Misrach 1990: 96)

It is therefore no surprise that public safety is a concern in the construction and use of Bravo 20. The museum would use state-of-the-art technology to defuse live munitions, putting the visitor on alert that there can be no guarantee of safety. This serves as another opportunity to educate the public that risks cannot be eliminated. That the US military has contaminated the land to such an extraordinary degree, that there has been little concern for public safety on the local level in preparation for war on an international level, that national security has usurped the safety of individual citizens and entire communities. This is the military folly that Bravo 20 seeks to illuminate for visitors.

There will be driving and walking tours to illustrate the effects of war. These include a bus tour of the range along a road to be called Devastation Drive, a walking tour called the Boardwalk of the Bombs, and a hiking trail up Lone Rock, for a panoramic view. All necessary precautions will be taken to ensure visitors' safety, and people will be encouraged not to stray from the boardwalks and bus routes that have been specifically cleared of potential explosives. The boardwalks, modelled after those at Yellowstone National Park and the Hawaii Volcanoes National Park, will be constructed to protect the tourist from shrapnel, craters, and other difficult terrain!

Outside the museum, on the bombing range, will be a number of important site exhibits reachable either by car or by tour bus (decommissioned personnel carriers) and by foot via the Boardwalk of the Bombs. Pullouts along Devastation Drive will provide rest stops and picnic areas, as well as various displays at the laser range, archaeological grounds, and other sites. The network of walk-ways will lead the visitor to a number of exhibits unreachable by vehicle…Each exhibit will have a site plaque describing its significance.

(Misrach 1990: 96)

It is also recognized that the site could bring in valuable tourist dollars, as military spending declines in Nevada. But the main driving force, far from monetary value, is the desire to show the effects of war, in this case a war on the environment and local people's health and wellbeing in the name of national security.

Conclusion

The devastation wrought by the splitting of the atom, brought to us by human ingenuity, defies all ways of knowing. All previous reality can be said to have been shattered in seconds on 6 and 9 August 1945, when nuclear weapons were used against Hiroshima and Nagasaki. Yet they have also been used, though cleverly presented as a means to prevent war, against the people of Nevada, of the Marshall Islands, of Kazakhstan, and Christmas Island. Atmospheric nuclear tests ensured that the burden of radioactive contamination be borne by all living beings on our planet. As the Hibakusha (the Japanese bomb survivors) remind us, 'we are all affected by the instruments of nuclear war'. We all live downwind and downstream. And the reach of the nuclear shadow extends ever outwards, through the continued use and abuse of nuclear power and the continued development of nuclear weapons.

The Sellafield Visitors' Centre and Bravo 20 are both part of public education campaigns: one where a history of bomb manufacturing is hidden under the mantle of nuclear power, and the other where a war on the environment, in preparation for a potential global omnicide, is exposed to youth in an effort to engage critical inquiry. At Bravo 20 children would be guided via covered walkways through a national sacrifice zone of the American West, signifying the vast open space as an ungainly lamb on the altar of the nuclear bomb.

Places have died for nuclear bombs, and people die because of them. Against the backdrop of fear perpetrated by the 'war on terrorism' – a war in which the US and UK military have mooted the idea of using 'mini-nukes' against non-nuclear nations or stateless actors – nuclear tourist traps can be places to play in which people and places have died and will continue to die into an untold future.[8]

Notes

1   No discussion of contemporary nuclear tourism is complete without mention of the Smithsonian Air and Space Museum. Bird and Lifschultz (1998) describe the Smithsonian Museum's attempt to commemorate critically the 50th anniversary of the bombing of Hiroshima and Nagasaki. Why the decision to use nuclear weapons against civilians was taken, and by whom, is still contested. The virtual cancellation of the Smithsonian exhibit (with the exception of the fuselage of the Superfortress *Enola Gay*, the airplane used to target Hiroshima) denied an opportunity for the public and political leaders to examine the history of nuclear weapons development and use. On 15 December 2003, the Smithsonian again opened a display of the *Enola Gay*, having been newly reassembled after 300,000 staff hours of restoration. As five years previously, there

was no analysis of the meaning of nuclear weapons, or history of the decision to use them. In this exhibit as with the last, the focus is on the Superfortress, and its transformation of war-fighting capabilities. See *www.nasm.si.edu/events/pressroom/releases/081803.htm*

2   On the internet at *www.atomictourist.com*

3   To recognize nuclear 'weapons' as being beyond the scope of conventional war materiel is essential to understanding their unique destructive capabilities. Rosalie Bertell conceived the notion of 'omnicide' to distinguish nuclear 'weapons' from their distant relatives on the battlefield in the family of war. She defines omnicide as: 'the concept of species annihilation...deliberately induced to end history, culture, science, biological reproduction and memory...It is the ultimate human rejection of the gift of life' (1985: 2).

The most popularly identified nuclear tourist sites, besides Hiroshima and Nagasaki, include the Los Alamos nuclear laboratories, and the Trinity Site, both located in New Mexico (for the more adventurous tourists, the abandoned nuclear test zones of the South Pacific offer opportunities for underwater exploration. See Davis 1994). Los Alamos was home to the Manhattan Project, a top-secret programme to develop the first nuclear weapons. Still an active nuclear weapons laboratory (see *www.wslfweb.org*), Los Alamos also hosts its own 'nuclear museum' espousing the 'peaceful' role of nuclear deterrence. The Trinity Site, where the first explosion of a nuclear device occurred in Alamogordo, New Mexico on 16 July 1945, is located on the White Sands Missile Base, and is owned by the US military. It is only open to tourists twice a year. The limited time-frame seems to add to its appeal, as it attracts visitors from all over the world. All that stands on the site is a conical-shaped, stone structure which bears a plaque, announcing that the nuclear age began in that precise location. Visitors are forbidden to take 'trinitite', the 'rock' that was 'manufactured' in the blast, which consists of fused sand resulting from the explosion's excessive temperatures. Still, trinitite is smuggled out and coveted by nuclear tourists who, by owning a 'sign' of the nuclear age, possess some part of that history.

4   At many nuclear reactor sites, tourism is encouraged, and 'Information Centres' have been established adjacent to power stations in the United States, the United Kingdom, France, Japan, Germany, and other nations.

5   Urry describes 'surrogate parents' of tourism as the travel agents, couriers, hotel managers, and others who 'relieve the tourist of responsibility' and protect them from 'harsh reality' (1990: 7). For the nuclear tourist this position is filled by the public relations personnel, museum guides, and the texts of brochures and exhibition displays, directing the tourist gaze.

6   The statistics cited were provided by staff at the Visitors' Centre on 8 October 2003.

7   Both in Britain and the United States there have been dialogue groups set up for concerned citizens to be a part of decision-making processes. The author participated in one such stakeholder group in the United States, and observed their proceedings in Britain, and believes that to say they represent a partnership between citizens, government, and energy corporations would be disingenuous. Although this trend may be changing, and it is a step forward to be given a place at the table, the current status of stakeholder dialogues has produced little in the way of actual policy changes.

8   The following web sources were used: Bureau of Atomic Tourism, *www.atomictourist.com*; National Museum of Nuclear Science and History, *www.atomicmuseum.org*; Indian Point Energy Center, *www.safesecurevital.org*; Smithsonian National Air and Space Museum, *www.nasm.si.edu/events/pressroom/releases/081803.htm*; Western States Legal Foundation, *www.wslfweb.org*.

The following films are also relevant: Beaver, Chris *et al.* (1983) *Dark Circle*, San Francisco: Independent Documentary Group; Loader, Jayne *et al.* (1982) *Atomic Café*, New York: The Archives Project.

Chapter 18

Death in Venice[1]

John Urry

What he needed was a break, an interim existence, a means of passing time, other air and a new stock of blood, to make the summer tolerable and productive. Good, then, he would go on a journey.

(Thomas Mann, *Death in Venice* [1912] 1955: 12)

Henry James reported that the hotels in San Remo were 'filled with English and German consumptives'.

(quoted in Pemble 1987: 86)

The Bombs in Spain Fall Mainly on the Tourists

(http://slate.msn.com/id/112743/)

Playful places

The chapters in this book show that in many different ways tourist activities or performances cannot be separated from the places that are visited. Indeed places are not fixed and unchanging but depend upon what gets bodily performed within them by 'hosts' and especially by various kinds of 'guest'. Places to play are brought into being through systems of organized and/or informal tourist performance. These performances stem from and reproduce complex systems of networked mobilities that only contingently stabilize certain places as being fit 'for play'. Thus places are economically, politically, and culturally produced through the multiple mobilities of people, but also of capital, objects, signs, and information moving at rapid yet uneven speed across many borders. The mobilities of people comprise tourists, migrants, design professionals, asylum seekers, backpackers, business and professional travellers, students, and other young people 'travelling the world' for their OE (overseas experience). There are moreover many overlaps between 'tourism' and these other kinds of business, professional, and migratory movement.

We have seen how certain places are 'in play' in relationship to these multiple mobilities. Places are dynamic, moving around and not necessarily staying in one 'location'; we have even seen this in the case of apparently fixed 'islands' whose images have changed over the years. Places 'travel' within networks of human (and, as authors in this book have argued, non-human) agents, of photographs, sand, cameras, cars, souvenirs, paintings, surfboards, computer screens, paintings, Viking ships, and so on. These objects extend what humans are able to do, what performances of place are possible. And the resulting networks swirl around, increasingly fluidic, changing the fixity of place and bringing unexpected new places 'into' play.

Places are situated at different stages and locations within global flows – there are places that go 'with the flow' and those that are left 'behind' – and these designations, as we have seen, can rapidly change. Some places to play 'move' closer to various global centres (which are themselves in play), while others move further away from the global stage upon which towns, cities, islands, and countries appear to compete, to mobilize themselves as spectacle, to develop their brand, and to attract visitors. Places themselves are 'in performance' on this relentless global stage. Such places to play are often hugely complex, and there is no single linear experience as networks playfully criss-cross the world, bringing the curtain up and down upon place after place.

And these places to play, even beaches of paradise, are not all fun. They are haunted by strange combinations of play and purgatory, of life and death, of desire and disease. It is as though the pleasures of paradise can be too excessive, and they turn over to reveal an underbelly of danger, death, and disease. *Paradise News* (1991), as David Lodge's novel is entitled, is news of the danger and of the dying as well as the living, of 'Death in Venice' as well as of play and pleasure (see Mann [1912] 1955; Visconti 1971).[2] This chapter takes the theme of danger and death to reveal its opposite, of the life of, and within, places to play.

In examining this underbelly of danger and death, I first consider how places to play are places often thought of as places *to die for*, places that cannot be missed since they mark and may complete the good and fulfilled life. Second, places to play can dramatically become *places that die*, that as we have seen may get extruded from the 'global order', moving into the slow or dying lane. Third, many places to play have been *places of death*, where death was that place's *raison d'être*. Fourth, global places are often *places in which to die*, attracting the consumptive, crippled, dying, harbingers of infectious diseases, as well as those putting their body to dangerous extremes. Finally, *places to play are potential sites of death and destruction*, with tourists and terrorists checking in side-by-side. Places to play are places of gates, camps, and surveillance excluding suicide bombers while encouraging suicidal bungee jumpers to enter.

Places to die for

There are many places to die for – national and especially global icons that the whole world wants to see within their lifetimes, before death strikes. People make the journey and their eyes feast upon that place of the imagination, such as the Louvre (chapter 9), the Taj Mahal (chapter 10), Machu Picchu (chapter 7), the palm-lined Caribbean beach (chapter 2), or perhaps now Barcelona (chapter 12). When Freud finally got to Athens in his forties he found the Parthenon he had known about from his childhood literally took his breath away. He could not believe that it really existed and he was there in its overwhelming presence (Rojek 1997: 56–57). Many tourists too often find it hard to believe that they are in the presence of that particular 'wonder of the world' they have known about in their dreams, often from childhood.

This desire to be in 'other places' is deadly serious and stems from the shift in the material practices of everyday life that can be over-simply characterized as moving from the social practices of *land* to *landscape* (Milton 1993). *Land* is a physical, tangible resource to be ploughed, sown, grazed, and built upon, a place of functional work. Land is bought and sold, inherited and left to children. To dwell is to participate in a life where productive and unproductive activities resonate with each other and with tracts of land, whose history and geography are known in detail. There is a lack of distance between people and things. Emotions are intimately tied into place, as Sarah Hall describes in *Haweswater*, set in the now-dead Lake District village of Mardale, submerged beneath the titular lake (2002; see Ingold 2000).

With *landscape*, however, there is an intangible resource whose key feature is appearance or look. This notion developed in western Europe from the eighteenth century onwards, part of the more general emergence of a specialized *visual* sense that differentiated itself from the other senses and came to be based upon novel technologies of the eye (see Febvre 1982). Areas of often wild, barren nature that had been sources of death, terror, and fear were transformed into landscape, places waiting at a distance for consumption by those dying to visit. Land became what Raymond Williams terms 'scenery, landscape, image, fresh air' (1972: 160; Bærenholdt *et al.* 2004: chapter 8).

The notion of landscape prescribes a visual structure of desire to the experiences of place. As Miss Bartlett paradigmatically declares in Forster's *A Room with a View*: 'A view? Oh a view! How delightful a view is!' – well worth waiting for, a view to die for (Forster [1908] 1955: 8). Maoris in New Zealand apparently never developed such a notion of landscape, never painting or indeed seeing their country as 'landscape' (Bell and Lyall 2002: 25).

In nineteenth-century northern Europe the desire for and capacity to fix places of the 'other' dramatically developed. Places came to be 'kodakized' as we might now say (West 2000; Bærenholdt *et al.* 2004: chapters 5 and 6). Such places of desire and fixing through the objects of the camera, tripod, and photograph included the Mediterranean (Pemble 1987), the Alps (Ring 2000), the Caribbean (Sheller 2003a), the Grand Canyon (Newmann 2002), the exotic Nile (Gregory 1999), stinking fishing villages (Lübbren 2001), and water generally (Anderson and Tabb 2002). According to Ruskin, the art of seeing detail through drawing was lost during the nineteenth century and was replaced by landscape fixed through photography that he, even as early as 1845, found fascinating (de Botton 2002: chapter 8). Other visual technologies of landscape, other actants within a 'landscape network', included the claude glass, the camera obscura, guidebooks, routes, sketching, the balcony, the railway carriage, and maps (Ousby 1990; Osborne 2000).

Buildings also constitute elements of such networks, they can make a place worth dying to see, to 'see Venice and die'. Machu Picchu or the Taj Mahal are utterly central to place, overwhelming signifiers. Machu Picchu signifies Inca heritage, the Taj the 'exotic orient'. New landmark hotels, office blocks, and galleries built by celebrity architects in Hong Kong, Barcelona, and Dubai are global icons, with the whole world increasingly now watching (chapters 11, 12, and 16).

Also, places are worth dying for if they are centres of intense or excessive conviviality where other people's performances produce liveliness or carnival, or movement to that place (see Urry 2002a, on the collective tourist gaze). Many moving people with the appropriate habitus indicate that this is *the* place to be, a place to die for, a place that cannot be missed, a place of life. The performances of moving, viewing others are obligatory for emotionally experiencing a place, such as cosmopolitan pre-SARS Hong Kong, the lively Caribbean, the Barcelona Olympics, climbing to the spiritual Machu Picchu, and so on. Baudelaire's notion of *flânerie* captures this emotional moving performance of place: 'dwelling in the throng, in the ebb and flow, the bustle, the fleeting' (quoted in Tester 1994: 2). There is a rich cacophony of time, some 'places

[are] full of time' and this gives them a distinct atmosphere, making them brim with 'cosmopolitan opportunity' (see Sennett 1991: chapter 7; Bærenholdt *et al.* 2004: chapter 3, on researching the 'atmosphere' of place; chapter 13 here on the intangible threatening atmosphere of a suburban estate).

Overall Buzard argues in *The Beaten Track* that Wordsworth's *The Brother* 'signifies the beginning of modernity…a time when one stops belonging to a culture and can only tour it', in order to compare, contrast, and collect, to see Venice and die (1993: 27; Barrell 1972). Some places come to be known as global icons, wonders of the world, places that can take the breath away, worth dying to see, to linger in, to be 'cosmopolitan' because one has been there to see for oneself. Touring the world is how the world is performed, as a connoisseur of places. Schultz indeed describes the astonishing '1000 Places to See Before You Die' (2003).

Places that die

In this 'touring' world, places come and go, some speed up and others slow down. A specialized visual sense characterizes this modern world. As E.M. Forster wrote: 'Under cosmopolitanism…Trees and meadows and mountains will only be a spectacle…', that is, landscape, not land (1931: 243). Many places are thus cosmopolitan, detached from nature and the environment and comprising, for their visitors, *images* or *signs* of trees, meadows, and mountains. Places are known about, compared, evaluated, possessed, but not 'dwelt within'. There is a tendency for *all* places on the global stage to become cosmopolitan and nomadic, known by the connoisseur whether of good clubs, views, walks, historic remains, landmark buildings, genuine *favelas*, and so on.

A place is thus not so much a place with its own associations and meanings for those dwelling or even visiting. Place is more a *combination* of abstract characteristics that mark it as more scenic or cosmopolitan or cool or exotic or global or environmentally degraded than other places. This language of abstract characteristics is a language of mobility, the expression of the life-world of mobile groups, of tourists, conference travellers, business people, or environmentalists. This is a consumption of movement, of bodies, images, and information, moving over, under and across the globe and subjecting it to abstract characterizations (see Szerszynski and Urry 2002).

Such movement provides new symbolic capital in class wars (see Mowforth and Munt 2003: chapter 5). Those mobilities, a 'fluid modernity', produce judgements of taste that feed *inter alia* into the development of ecotourist resorts (chapter 4). According to Mowforth and Munt, 'trendies are on the trail' developing new symbolic capital as they circumnavigate the world and presume to save it through being 'ecotourists' in Belize, walking at high altitude through the Andes or performing round-the-world travelling (2003; see chapters 4, 7, and 15 above).

Place after place gets consumed. Places are consumed, wasted, and used up as they are 'toured', becoming only, in the end, a set of abstract characteristics. Tourist practices move on, as we saw in the case of Cyprus, and leave behind those places no longer desired, no longer to be used, where few still come to play. People may try to escape from where there is a 'drudgery of place', from being inexorably tied there, where time is fixed and unchanging. Such places remain heavy with time and are left behind in the 'slow lane', as with many old-style seaside resorts in northern Europe and North America. They may be reincarnated as places for the performances of nostalgia. The Isle of Man advertises: 'You'll look forward to going back', experiencing a seaside holiday as remembered from one's distant childhood (Urry 2002a: chapter 5; Frow 1997).

Places to play are fundamentally places of service and material consumption. Through consuming certain goods or services the place is experienced/consumed (Urry 1995). Goods or services are metonymic of the place, with the part standing for the whole. The consuming of place normally involves goods or services that are unique or culturally specific to that place. As Molotch says: 'touristic stuff has a more definite placeness' (2002: 677). Products can be folded into place and places come to be as they are through multiple consumed products. People eat, drink, collect, gamble, scuba dive, surf, bungee jump the 'other' (Urry 2002a: 3).

Sometimes such places to play are places of excess, where consumption (though not always material consumption) is performed to dangerous extremes, so differentiating that place on the global stage. Examples of such consumption to excess include lengthy high-level walking and running in the Andes (chapter 7), recreational drug use in Cyprus (chapter 3), and the multiple dangerous pleasures on a Caribbean beach of paradise (chapter 2).

But places are also often full of bitter disappointment, frustration, and despair, where performances cannot be realized. This is paradigmatically captured in Alex Garland's book (1997) and film (2000) of *The Beach* (see also Campbell 1987). The contrast between people's fantasies of a place and the performances it actually affords are a constant trope in tourist tales. Thus we have seen examples of places where the objects of performed consumption are not available (parts of Cyprus closed for the winter), the service quality is degraded compared with the destination's place-image (as in 'English' Harrogate or cosmopolitan Barcelona), the place-to-die-for is overrun with cheap souvenir stalls (the Taj Mahal), ecotourists discover the pristine coral reef has been destroyed by earlier mass tourism, or the middle-aged find their playplace is over-run by promiscuous, drunk, drug-using 'party-goers' producing a one-generation (18–30) 'party-space' (as at Ayia Napa or Barcelona).

Where the place no longer affords those performances of play that visitors are seeking, then it may be in its death throes – a place of degraded consumption. More generally, the place of play where no-one much still plays seems more dead than ever (such as Tatilya on the edge of Istanbul).

Playplaces of death

But there is something even stranger about certain places to play. Consider some places of pleasure: the grassy knoll in Dallas, Changi Jail in Singapore, Nazi Occupation sites in the Channel Islands, Gracelands, Dachau, dangerous coal mines, war memorials, Hiroshima (see figure 18.1), slave plantations, Northern Ireland, Northern Cyprus under Turkish 'occupation', west African slave forts, Pearl Harbor, Robben Island in South Africa, Sarajevo's 'massacre trail', Jim Morrison's grave in Paris, Titanic, and Auschwitz-Birkenau (a UNESCO World Heritage Site). These places to play are all places of violent death, of many or of a single icon (such as Diana, JFK, James Dean, Sharon Tate; see Lennon and Foley 2000). It has been suggested that 'war' sites constitute the largest category of tourist destinations (Sharpley 2003b: 3).

Figure 18.1   Destroyed Hiroshima now a place for tourist pilgrimage

In this book we have examined how two apparently bloodthirsty cultures have been turned into cultures to consume, and related places in which to play and in part to perform – as Vikings (chapter 8) or as Incas (chapter 7). We also examined the strange phenomenon of nuclear tourism where visitors are brought into dangerous proximity to radioactive materials that can be made to appear playful and pleasurable, especially to young children visiting (chapter 17, and see figure 17.1).

There are many examples of 'dark tourism', to use Lennon and Foley's term (2000), or 'holidays in hell' (O'Rourke 1989), 'black spots' tourism (Rojek 1993), or holidays in the 'danger zone' (Adams 2004). These places of awesome death, massive danger, or untold suffering have contingently come to be performed as places of leisure. They often charge an entrance fee, provide interpretation, and sell various other services. And this turning of death into public play often results from mediatization – the simultaneous reporting and witnessing around the world of global suffering – with tourists as increasingly significant witnesses to that suffering.

Such places of death, although for play, are not necessarily Disneyfied or McDonaldized. This is because many such places have been developed and continue because of well-organized enthusiasts and fans (Hoggett and Bishop 1986; Abercrombie and Longhurst 1998: chapter 5). Such enthusiasts perform work, normally involving reciprocity and mutual aid. Much emphasis is placed upon acquiring arcane forms of knowledge about the atrocity, place, or person connected to the site. And there is often strong resistance to commodification. Such enthusiasts seek to keep 'alive' the memories of their particular race, religion, star, culture, or peoples. Organized fans or enthusiasts bring this experience of death into the public eye, to make the world witness it collectively through a public memorial.

But the bringing into public witnessing is normally realized through making that place open to tourists. Collective witnessing occurs through tourists coming and paying 'their respects' in public. It is as though the visits of tourists, as pilgrims to sacred sites of death, can produce an immortality of those that died in that place or for that cause. So these places entail complex performances: for tourists to grieve in public at the death of a race, a nation, a leader, or a star known normally through mediatized images.

These places of death and play are complex, entailing performances of respect, collective grief, and emotion. As a prominent notice instructs visitors at the Arlington National Cemetery where Kennedy is buried: 'Silence and Respect' (Lennon and Foley 2000: 88). That is how Arlington is to be performed by tourists.[3]

Playing with death

Places to play are often full of the ill, the dying, and those seeking to confront bodily their own immortality. We noted in chapter 6 the development in western Europe, during the eighteenth and nineteenth centuries, of spa towns such as Harrogate, Wiesbaden, Vichy, Bath, and Baden-Baden (Blackbourn 2002). Taking the waters and being immersed within them promised recovery to the ill and the dying. These were tourist sites in which to perform recovery from crippling illnesses.

European tourism began with the provision of services for the ill, with a high proportion of visitors suffering recognizable symptoms. 'Taking the waters' became a fashionable performed therapy within various medical regimes. Water, internal and external, was seen as the antidote to the multiple diseases of civilization (on water and leisure, see Anderson and Tabb 2002). Subsequently, spas developed into places of luxurious pampered pleasure (see Switzer 2002 on these shifts in Budapest).

Such spa towns were socially select places with entry limited to those who could afford housing in the town (as reported in Jane Austen's novels) or later to those who could afford the limited number of expensive hotels. Most spa towns have retained a place-image as socially select, often hiding their many workers from view. These were places where a cosmopolitan élite, increasingly able to travel by train, gathered from across Europe, and there was a growing circuit of travel between these fashionable places, drawing them 'closer' together. These spas provided cultural capital, enabling taste-setters to indulge in consumer practices in such places of services, places of play for the ill or 'Disneylands for the upper classes' (Blackbourn 2002: 15; and see chapter 6).

Early seaside resorts also developed as places of medical treatment. The beach was initially a medical zone, as Shields describes in the case of Brighton (1991). Only later did beaches turn into a zone of pleasure and especially a place for play, a paradise on earth (first in the Mediterranean, and later paradigmatically in the Caribbean: see chapter 2). In the eighteenth and early nineteenth centuries, beaches were places where the ill and the infirmed were 'dipped' into the sea because of its presumed health-generating properties. 'Dippers' – large, apparently sexless mature women – provided this distinctly performed 'service'. Beaches are often populated with the infirmed seeking to perform recovery and convalescence. In *Death in Venice*, the ageing and disappointed writer Aschenbach dies in his deckchair on the beach as the object of his desire walks out into the sea (Mann [1912] 1955; Visconti 1971). Throughout the book and the film there are many echoes of death and mortality, with one early scene occurring on the steps of Funeral Hall. The Caribbean too was seen as a place for the infirm. Cuban air was strongly recommended for treating tuberculosis as early as the 1830s, while Jamaica in 1903 was described as a 'veritable Mecca for the invalid' (Sheller 2003a: 65).

Many seaside resorts have remained as places for the ill to take the waters and the air, to receive treatment and to convalesce. Seaside resorts often have high concentrations of nursing homes, the retired and the infirm, as medical and fitness tourism has developed on a widespread scale. Cuba has an interesting comparative advantage in contemporary medical tourism because of the legacy of its good Communist health service (chapter 2).

But *Death in Venice* is also a story of infection, as Asiatic cholera sweeps through Venice. There are many connections between the mobilities of people and of illness. Cresswell brings out the connections between the threat of syphilis and new technologies of mobility (2000). High rates of international mobility have generated new risks, such as AIDS or SARS, which are both diseases of contemporary mobilities – making them distinctively modern plagues according to Farmer's definition (1999). In the recent history of Hong Kong we saw how the pattern of spread of SARS resulted from particular patterns of travel within the Chinese diaspora (chapter 11).

And typically, as Thomas Mann writes in *Death in Venice*: 'the city was not swayed by highminded motives or regard for international agreements. The authorities were more actuated by fear of being out of pocket...by apprehension of the large losses the hotels and shops that catered to foreigners would suffer in case of panic and blockade' ([1912] 1955: 72). Thus the threat of death to the place of play, to its pathologization, can lead to more deaths of visitors and locals as the mobile risks are swept under the metaphorical carpet. Places to play are immensely vulnerable to the movements of illnesses and of the fear of illness, which can overnight turn a place to play into a place fearing death. Panic can cause visitors to shun that place, revealing the fragility of most places to play (as with SARS in Hong Kong: chapter 11).

But tourism and travel does not always involve the minimizing of risk. It seems that part of the allure of the Caribbean is that 'danger' is just around the corner, just beneath the veneer of the playplaces. Tales of pirates, Rastas, drugs, and Yardies all contribute to performing 'dangerous tourism' in the paradise islands (chapter 2; Sheller 2003a). More recent tourism performances also involve putting the body into other kinds of personal danger; Sennett says that 'the body comes to life when coping with difficulty' (1994: 310; Macnaghten and Urry 2001). There are many guidebooks now for 'dangerous travel' (Schroeder 2002: 73) as well as a recent BBC TV series on *Holidays in the Danger Zone* (see Adams 2004 on danger-zone tourism).

Extreme sports have developed as new versions of the tourist gaze, involving distinctly dangerous and mobile tourist performances. These performances of bodily extremes include bungee jumping, off-piste skiing, paragliding, base jumping, skydiving, whitewater rafting, and high-altitude walking. Space travel seems to be the new final frontier for the adventurous tourist, with a quite astonishing array of companies, governments, and other organizations competing to produce a stable network that will enable 'mass' space travel to occur within a decade or so (see Cater 2003).

Thomas Mann once wrote that modernity, and especially those 'touring' this world, are in love with the abyss (quoted in Bell and Lyall 2002: 23). New Zealand especially has developed new performances of sublime experiences of the abyss. There, according to Bell and Lyall, is an accelerating sense of the sublime where 'nature provides a site in which tourists indulge their dreams of mastery over the earth; of being adventure heroes starring in their own movies' as they seek to cheat death (2002: 22). New Zealand has become that place where the 'glorious vista' provides appropriate locations for the dynamic consumption of the 'accelerated sublime' that increasingly is 'the way that New Zealand packages landscape for consumption' (Bell and Lyall 2002: 36). High-altitude walking in the Andes similarly provides a site for visitors to challenge their own immortality and to approach Machu Picchu through a 'proper' (self-powered) performance, just like the Incas would have done on a four-day high-altitude trek (chapter 7)!

Tourism and terrorism in places to play

Some tourist places have always been places of danger, of course, where crime and fears around personal safety are central to the places to which tourists flock. Rio de Janeiro is a key example of this hyper-concentration of tourism and criminality, where criminals from the *favelas* target tourists who provide a honeypot (chapter 14). There are many other examples of the attraction of tourists for criminals, for mugging, prostitution, pickpocketing, and various illegal trades. The campaign of ETA terrorists in the Basque country especially targeted tourist areas, using bombs directed against tourists as a key part of their campaign to secure Basque independence.

But something new is now on the agenda, with the transformation of terrorism away from bomb campaigns directed against specific targets in order to effect regime change within a society (as with the IRA in Northern Ireland). Now terrorism has gone global – there is the 'revenge of the nomads' (Bauman 2002: 13). Global terrorism seeks to challenge the global power of the United States (whether viewed from an Islamist perspective as a satanic power, or from an anti-colonialist one as a neo-colonialist) and its allies. In this new world disorder, places that attract Western tourists are the new target. Tourists are in the front line of the new global warfare. Places to play are where death is just round the corner (Cairo, Luxor, New York, Bali, Mombasa, Jakarta, Kashmir), attracting deadly visits from those seeking the mass sacrificial deaths of others. Potential death *and* the fear of death stalk almost all the places to play that are examined in this book.

Such places increasingly attract tourists *and* terrorists. Some of the time terrorists are tourists, only intermittently transmuting into terrorists. And the weapon of the weak is of fear, to induce panic into those 'innocent tourists' playing away, performing what they are meant to perform in places of play. 'The new fear is bound up with radical uncertainty. Terror hits randomly...the new terror is blind and diffuse...highly invisible' (Diken and Lautsten 2004: 2). The new fear is like an epidemic, potentially striking at the airport, on the plane, at the hotel, in the nightclub, on the beach, at the petrol station, on the tourist bus, in the apartment. To be a tourist is to be in the front line, to play is potentially to die. In the aftermath of the iconic event in this new world disorder, the terrorist attacks on New York and Washington of 11 September 2001, there were calls for American 'patriotic tourism', to make sure that they got on planes and went to places to play in order to show the enemy that the fear of death could be defeated and the terrorists could not win. But in some sense at least, 'Bin Laden has already won; his victory consists of creating an all-consuming fear' (Diken and Lautsten 2004: 14).

There are two further points. First, with the passing of time, places of death become new places to play. So 'Ground Zero' in Manhattan, or the Falls and Shankhill Roads in Belfast, are now on the tourist map, waiting for visitors to come. 'Ground Zero' attracts more visitors than previously came to see the view from the top of the Twin Towers (Sharpley 2003b: 2), while in Belfast, visitors can take a 'Troubles Tour'. Places of death routinely transmute into places for visitors, appearing on ever-new tourist itineraries, part of the performances of dark tourism. We saw something similar in the development of visits to the dangerous suburbs around Paris (chapter 13) as well as *favela* tours to experience the 'real' Rio of drugs and danger (chapter 14). In such cases visitors develop distinct ways of performing their itinerary safely (or as safely as possible in an inherently dangerous situation) in an environment lacking typical sites and objects that enable tourist performances.

Second, this new invisible enemy encourages new forms of control. International tourists need exceptionally refined systems of surveillance in order to keep them on the move across the globe. In the United States this requirement provoked an unprecedented event, the nationalizing of airport security and the general development of control systems over the 550 million people who enter the US every year (Diken and Lautsten 2004: 3). The notion of 'inside' and 'outside' erodes; all are inside and outside simultaneously. Power, play, and terror are everywhere.

Moreover, because of the fluid nature of global media images, the death of tourists in the current global order is perceived in the United States as an attack on the United States. This spreads the contagion of fear, the call for more patriotic tourism to combat the invisibly mutating virus of al-Qaida, and the proliferation of new forms of remote-control security systems (see Blake and Thea Sinclair 2003 on the travel effects of 11 September).

Indeed tourists are now subject to the most intrusive regulation; places to play increasingly mimic airports, using many of the same kinds of monitoring, surveillance, and regulation techniques of what is increasingly called the 'frisk society'. In order to perform in places of play, to be a consumer in the global marketplace, tourists are subject to powerful and extensive systems of monitoring and regulation by corporations and states (see Watson 2003). It is the 'soft targets' of people playing in those places that are in the front line of the war on terror; they are, we might say, global warriors. And as both terrorists and tourists are 'on the move' and yet have to be kept 'apart', so gates, camps, sniffer dogs, cameras, face recognition biometric cameras, smart cards, iris recognition, satellites, listening bugs, and Total Information Awareness are all part of the performances of contemporary travel and tourism. In order that we can enter paradise for a week, systems of personal security are morphing into a Big Brother that would have George Orwell turning in his grave. This is the latest dystopian 'news from paradise' in a world of global terror/tourism. Big Brother is watching, especially when we seek to 'get away from it all'.

Notes

1   I am very grateful for the comments of Mimi Sheller on this chapter, as well as those from the other contributors to this book.
2   See Minca and Oakes 2004a, for some further travels in Venice.
3   See Sharpley 2003b for an extensive discussion of dark tourism that links various forms to a typology of different consumption practices.

Bibliography

Abercrombie, N., and Longhurst, B. (1998), *Audiences*, London: Sage.

Acheson, J. (1988), *The Lobster Gangs of Maine*, Hanover, NH: University Press of New England.

Adam, B. (1998), *Timescapes of Modernity: The Environment and Invisible Hazards*, London: Routledge.

Adams, K. (2004), 'Terror and Tourism: Charting the Ambivalent Allure of the Urban Jungle', in Minca, C., and Oakes, T. (eds), *Tourism and the Paradox of Modernity*, Minneapolis: University of Minnesota Press.

Addyman, P.V. (1994), 'Reconstruction and Interpretation: The Example of the Jorvik Viking Centre', in Gathercole, P., and Lowenthal, D. (eds), *The Politics of the Past*, London: Routledge.

Adriansen, I. (2003), *Nationale Symboler i Det Danske Rige 1830–2000* [National Symbols in the Danish Realm 1830–2000], Copenhagen: Museum Tusculanum.

Ahmed, S., Castañeda, C., Fortier, A., and Sheller, M. (eds) (2003), *Uprootings/Regroundings: Questions of Home and Migration*, Oxford: Berg.

Aitkenhead, D. (2000), 'Pleasure Island', *Guardian*, Travel Section, 30 November: 2–3.

Akis, S., Peristianis, N., and Warner, J. (1996), 'Resident Attitudes to Tourism Development: The Case of Cyprus', *Tourism Management* 17: 481–94.

Albertsen, N. (1999), 'Urbane atmosfærer' [Urban Atmospheres], *Sosiologi i dag* 4: 5–29.

Alexander, M.J. (1997), 'Erotic Autonomy as a Politics of Decolonization: An Anatomy of Feminist and State Practices in the Bahamas Tourist Economy', in Alexander, M.J., and Mohanty, C.T. (eds), *Feminist Genealogies, Colonial Legacies, Democratic Futures*, New York: Routledge.

Amin, A., and Thrift, N. (2002), *Cities: Reimagining the Urban*, Cambridge: Polity Press.

Anderson, S., and Tabb, B. (eds) (2002), *Water, Leisure and Culture*, Oxford: Berg.

Andronikou, A. (1987), *Development of Tourism in Cyprus: Harmonisation of Tourism with the Environment*, Nicosia: Cosmos.

Anglo-Saxon Chronicle (2000), translated and edited Michael Swanton, London: Weidenfeld & Nicolson.

Apostolopoulos, Y., and Gayle, D. (eds) (2000), *Island Tourism and Sustainable Development: Caribbean, Pacific and Mediterranean Experiences*, Westport, CT: Praeger.

Appadurai, A. (1990), 'Disjuncture and Difference in the Global Cultural Economy', in Featherstone, M. (ed.), *Global Culture*, London: Sage.

—— (1996), *Modernity at Large: Cultural Dimensions of Globalization*, Minneapolis: University of Minnesota Press.

—— (2000), 'Disjuncture and Difference in the Global Cultural Economy', in Held, D., and McGrew, A. (eds), *The Global Transformations Reader*, Oxford: Blackwell.

Arellano, A. (1997), *All Cuzco*, Barcelona: EDO.

—— (2000), 'Echando una Mirada al Tema del Patrimonio', *Turismo y Patrimonio* 2: 101–14.

—— (2003), 'The Inca Heritage: Indigenism, Authenticities, and Tourisms', Ph.D. thesis, Lancaster University.

Asplund, J. (1987), *Det Sociala Livets Elementära Former*, Gothenburg: Korpen.

Attalides, M. (1979), *Cyprus: Nationalism and International Politics*, New York: St Martin's Press.

Audrerie, D., Souchier, R., and Vilar, L. (1998), *Le patrimoine mondial*, Paris: PUF.

Augé, M. (1995), *Non-Places: Introduction to an Anthropology of Supermodernity*, London: Verso.

Ayers, R. (2000), 'Tourism as a Passport to Development in Small States: The Case of Cyprus', *International Journal of Social Economics* 27: 114–33.

Bærenholdt, J.O. (2001), 'Greenland on the Margin: The Nordic, Tourist and Educational Geographies of Greenland', *Nordisk Samhällsgeografisk Tidsskrift* 34: 31–50.

Bærenholdt, J.O., Haldrup, M., Larsen, J., and Urry, J. (2004), *Performing Tourist Places*, Aldershot: Ashgate.

Ballantine, J.L., and Eagles, P.F.J. (1994), 'Defining Canadian Ecotourists', *Journal of Sustainable Tourism* 2: 210–14.

Barber, S. (2001), *Extreme Europe*, London: Reaktion Books.

Barke, M., and Towner, J. (1996), 'Urban Tourism in Spain', in Barke, M., Towner, J., and Newton, M.T. (eds), *Tourism in Spain: Critical Issues*, Wallingford, Oxfordshire: C.A.B. International.

Barker, E. (1999), 'Exhibiting the Canon: The Blockbuster Show', in Barker, E. (ed.), *Contemporary Cultures of Display*, London: Yale University Press in association with the Open University.

Barrell, J. (1972), *The Idea of Landscape and the Sense of Place, 1730–1840*, Cambridge: Cambridge University Press.

Bartolucci, M. (1996), 'Barcelona: The Current Construction Slowdown...', *Metropolis* 16: 60–97.

Barton, L. (2003), 'Come On In, the Water is Lovely', *Guardian*, Travel Section, 10 May; available online at *http://travel.guardian.co.uk/news/story/0,7445,952675,00.html*

Basch, L., Glick Schiller, N., and Szanton Blanc, C. (1994), *Nations Unbound: Transnational Projects, Postcolonial Predicaments, and Deterritorialized Nation-States*, Amsterdam: Gordon & Breach.

Bassett, C., and Wilbert, C. (1999), 'Where do you Want to Go Today? (Like It or Not): Leisure Practices in Cyberspace', in Crouch, D. (ed.), *Leisure/Tourism Geographies: Practices and Geographical Knowledge*, London: Routledge.

Bastin, R. (1984), 'Small Island Tourism: Development or Dependency?', *Development Policy Review* 2: 79–90.

Bauman, Z. (1995), *Life in Fragments: Essays in Postmodern Morality*, Oxford: Blackwell.

—— (2000), *Liquid Modernity*, Cambridge: Polity Press.

—— (2002), *Society Under Siege*, Cambridge: Polity Press.

Bazin, G. (1967), *The Museum Age*, Brussels: Desoer.

Beach Boys (1961), 'Let's Go Trippin' ', Los Angeles: Guild Music.

Bech, H. (1996), '(Tele)urban Eroticisms', *Parallax* 1: 89–100.

Beck, U. (1992), *Risk Society: Towards a New Modernity*, London: Sage.

—— (1995), *Ecological Politics in an Age of Risk*, Cambridge: Polity Press.

Beder, S. (1991), 'Controversy and Closure: Sydney's Beaches in Crisis', *Social Studies of Science* 21: 223–56.

Bell, C., and Lyall, J. (2002), 'The Accelerated Sublime: Thrill-Seeking Adventure Heroes in the Commodified Landscape', in Coleman, S., and Crang, M. (eds), *Tourism: Between Place and Performance*, New York: Berghahn.

Bellah, R. (1991), *Beyond Belief*, Berkeley: University of California Press.

Benedict, B. (1983), 'The Anthropology of World's Fairs', in Benedict, B. (ed.), *The Anthropology of World's Fairs*, London: Scolar Press.

Benedikt, M. (ed.) (1991), *Cyberspace: First Steps*, Cambridge, MA: M.I.T. Press.

Benítez Rojo, A. (1996) [1992], *The Repeating Island: The Caribbean and the Postmodern Perspective*, second edition, trans. J.E. Maraniss, Durham, NC: Duke University Press.

Benjamin, W. (1978), 'Paris, Capital of the Nineteenth Century', in Demetz, P. (ed.), *Reflections*, New York: Schocken.

Bennett, A. (1994), *Writing Home*, London: Faber & Faber.

Berman, M. (1988), *All That is Solid Melts into Air*, New York: Penguin.

Bertell, Rosalie. (1985) *No Immediate Danger: Prognosis for a Radioactive Earth*, London: The Women's Press.

Bianchini, F., and Parkinson, M. (1993), *Cultural Policy and Urban Regeneration*, Manchester: Manchester University Press.

Bird, K., and Lifschultz, L. (eds) (1998), *Hiroshima's Shadow: Writings on the Denial of History and the Smithsonian Controversy*, Stony Creek, CT: Pampleteer's Press.

Birkett, D. (2002), *Ethical Tourism*, London: Hodder & Stoughton.

Bishop, Elizabeth (1979), *The Complete Poems, 1927–1979*, New York: Farrar, Strauss, and Giroux.

Blackbourn, D. (2002), 'Fashionable Spa Towns in Nineteenth Century Europe', in Anderson, S., and Tabb, B. (eds), *Water, Leisure and Culture*, Oxford: Berg.

Blair, J. (1985), *The Illustrated Discography of Surf Music, 1961–1965*, Ann Arbor, MI: Pierian Press.

Blake, A., and Thea Sinclair, M. (2003), 'Tourism Crisis Management: US Response to September 11', *Annals of Tourism Research* 30: 813–32.

Bloom, W. (1991), *The New Age: An Anthology of Essential Writings*, London: Rider.

BNFL (British Nuclear Fuels Ltd) (1997), *Investigating Sellafield Visitors' Centre*, Cheshire: Resources for Learning

Boden, D., and Molotch, H. (1994), 'The Compulsion to Proximity', in Friedland, R., and Boden, D. (eds), *Nowhere: Space, Time and Modernity*, Berkeley: University of California Press

Böhme, G. (1995), *Atmosphäre: Essays zur neuen Ästhetik*, Frankfurt am Main: Suhrkamp.

Bolter, H. (1996), *Inside Sellafield: Taking the Lid off the World's Nuclear Dustbin*, London: Quartet.

Bolton, H.C. (1891), 'Some Hawaiian Pastimes', *The Journal of American Folklore* 4: 21–26.

Boo, E. (1990), *Ecotourism: The Potentials and the Pitfalls*, 2 vols, Washington, D.C.: WWF.

Booth, D. (1994), 'Surfing '60s: A Case Study in the History of Pleasure and Discipline', *Australian Historical Studies* 26: 262–79.

——— (1995), 'Ambiguities in Pleasure and Discipline: The Development of Competitive Surfing', *Journal of Sport History* 22:189–206.

Botton, A. de (2002), *The Art of Travel*, London: Hamish Hamilton.

Bottrill, C.G. (1995), 'Ecotourism: Towards a Key Elements Approach to Operationalizing the Concept', *Journal of Sustainable Tourism* 3: 45–54.

Bourdieu, P. (1984), *Distinction*, London: Routledge.

——— (1993), *The Field of Cultural Production*, London: Routledge & Kegan Paul.

Bourdieu, P., and Dabel, A. (1991), *The Love of Art*, London: Routledge.

Bowcott, O., (2001), 'Monumental Expense', *Guardian Unlimited*, 18 August; available online at *http://travel.guardian.co.uk/saturdaysection/story/0,8922,538336,00.html*

Briguglio, L., Archer, B., Jafari, J., and Wall, G. (eds) (1996), *Sustainable Tourism in Island and Small States: Issues and Policies*, London: Pinter.

Britton, S. (1991), 'Tourism, Capital and Place: Towards a Critical Geography of Tourism', *Environment and Planning D: Society and Space* 9: 451–78.

Brown, A. (2003), 'Cathay: No Plans to Ground Fleet', CNN, 13 April; available online at *http://edition.cnn.com/2003/HEALTH/04/13/sars.asia/*

Brown, P. (1998), 'Meacher Considers Closing Sellafield', *Guardian*, 18 July.

——— (2003), 'Sellafield Shutdown Ends the Nuclear Dream', *Guardian*, 26 August; available online at *http://www.guardian.co.uk/uk_news/story/0,3604,1029333,00.html*

Browning, B. (1998), *Infectious Rhythm: Metaphors of Contagion and the Spread of African Culture*, New York and London: Routledge.

Bryson, B. (1995), *Notes from a Small Island*, London: Black Swan.

Buck-Morss, S. (1989), *The Dialectics of Seeing: Walter Benjamin and the Arcades Project*, Cambridge, MA: M.I.T. Press.

Buzard, J. (1993), *The Beaten Track: European Tourism, Literature and the Ways to 'Culture', 1800–1918*, Oxford: Clarendon Press.

Cade, J. (1998), *Agatha Christie and the Eleven Missing Days*, London: Peter Owen.

Cadwalladr, C. (2001), 'No rest from the wicked', *The Daily Telegraph*, 2 June; available online at *www.telegraph.co.uk/travel/main.jhtml?xml=/travel/2001/06/02/etayia.xml*

Cairncross, F. (2001), *The Death of Distance 2.0: How the Communications Revolution Will Change Our Lives*, London: Texere.

Callois, R. (2001), *Man Play and Games*, Chicago: University of Chicago Press.

Cameron, N. (1991), *An Illustrated History of Hong Kong*, Oxford: Oxford University Press.

Campbell, C. (1987), *The Romantic Ethic and the Spirit of Modern Consumerism*, Oxford: Blackwell.

Camus, Albert [1946–49] (1978), *Journaux de Voyage*, Paris: Gallimard.

Carl's Garage Site (2003), 'Ayia Napa – Fantasy Island – The Replies?', *www.keconnect.co.uk/~carl/naparply.htm* (accessed 6 February 2004).

Carreras i Verdaguer, C. (1993), 'Barcelona 92: Una Politica Urbana Tradicional', *Estudios Geograficos* 212: 467–81.

Castells, M. (1996), *The Rise of the Network Society*, Oxford: Blackwell.

——— (2000), *The Rise of the Network Society*, second edition, Oxford: Blackwell.

Cater, C. (2003), 'The Final Frontier? Opportunities for Astrotourism', paper given to the *Taking Tourism to the Limits* Conference, Waikato University, Hamilton, 9–11 December.

Cavalcanti, M. (2001), 'Demolição, Batalha e Paz: favelas em manchetes', M.A. thesis, Rio de Janeiro: School of Communications, Federal University of Rio de Janeiro.

Chaloupka, W. (1992), *Knowing Nukes: The Politics and Culture of the Atom*, Minneapolis: University of Minnesota Press.

Chamberlain, M. (ed.) (1998), *Caribbean Migration: Globalized Identities*, London: Routledge.

Chandler, D. (1997), 'Writing Oneself in Cyberspace': *www.aber.ac.uk/media/Documents/short/homepgid.html* (accessed 6 February 2004).

Chaney, D. (2002), 'The Power of Metaphors in Tourism Theory', in Coleman, S., and Crang, M. (eds), *Tourism: Between Place and Performance*, Oxford: Berghahn.

Chaney, E. (1987), 'The Context of Caribbean Migration', in Sutton, C., and Chaney, E. (eds), *Caribbean Life in New York City: Sociocultural Dimensions*, New York: Center for Migration Studies.

Chang, W., and Yin, C. (1997), 'Exhibition Estimation and Visitor Survey of "the Golden Impression – Masterpieces of Musée d'Orsay"', *Journal of the National Museum of History*, Taipei: National Museum of History.

Chatterton, P., and Hollands, R. (2003), *Urban Nightscapes: Youth Cultures, Pleasure Spaces and Corporate Power*, London: Routledge.

Chen, W., Boase, J., and Wellman, B. (2002), 'The Global Villagers: Comparing Internet Users and Uses Around the World', in Wellman, B., and Haythornthwaite, C. (eds), *The Internet in Everyday Life*, Oxford: Blackwell.

Cheney, G.A. (1995), *Journey to Chernobyl: Encounters in a Radioactive Zone*, Chicago: Academy Chicago.

Cheung, C-F. (2003), 'Shanghai Visitor Sees Storm Blow Spending Spree Off Course', *South China Morning Post*, 3 September; available online at *http://hongkong.scmp.com/cgi-bin/gx.cgi/AppLogic+FT.../Printacopy&aid=ZZZX9GGSHJ* (paid subscription necessary).

Clark, S., and Gaile, G. (1998), *The Work of Cities*, Minneapolis: University of Minnesota Press.

Clifford, J. (1997), *Routes: Travel and Translation in the Late Twentieth Century*, Cambridge, MA: Harvard University Press.

Clift. S., and Carter, S. (eds) (1998), *Tourism and Sex: Culture, Commerce and Coercion*, London and New York: Pinter.

Clough, R. (1998), 'The Enduring Influence of the Republic of China on Taiwan Today', in Shambaugh, D. (ed.), *Contemporary Taiwan*, Oxford: Oxford University Press.

Club 18–30 (2003), 'Cyprus: information', *www.club18–30.co.uk/engine_005.asp* (accessed 6 February 2004).

Cohen, E. (1972), 'Towards a Sociology of International Tourism', *Social Research* 39: 164–82.

——— (1979) 'A Phenomenology of Tourist Experiences', *Sociology* 13: 179–201.

——— (1987) 'Alternative Tourism – a Critique', *Tourism Recreation Research* 12: 13–18.

——— (1988) 'Authenticity and Commoditization in Tourism', *Annals of Tourism Research* 15: 371–86.

Coleman, S., and Crang, M. (2002a), 'Grounded Tourists, Travelling Theory', in Coleman, S., and Crang, M. (eds), *Tourism: Between Place and Performance*, Oxford: Berghahn.

Coleman, S., and Crang, M. (eds) (2002b) *Tourism: Between Place and Performance*, Oxford: Berghahn.

Conlin, M., and Baum, T. (1995), 'Island Tourism: an Introduction', in Conlin, M., and Baum, T. (eds), *Island Tourism: Management Principles and Practice*, Chichester: John Wiley.

Cook, S., *et al.* (1998), *USA: The Rough Guide*, London: Rough Guides.

Cope, R. (2000), 'Republic of Cyprus', *Travel & Tourism Intelligence: Country Reports* 4: 3–21.

Corbin, A. (1994), *The Lure of the Sea: The Discovery of the Seaside in the Western World, 1750–1840*, Berkeley: University of California Press.

Cotgrove, S., and Duff, A. (1980), 'Environmentalism, Middle Class Radicalism and Politics', *Sociological Review* 28: 333–51.

Cowan, T. (2002), 'Hanford's B Reactor: An Icon of Atomic Age', *Seattle Post-Intelligencer*, 23 November.

CPC (1989), *Five Year Development Plan, 1989–1993*, Nicosia: Central Planning Commission, Planning Bureau.

Crang, M. (1994), 'On the Heritage Trail: Maps of and Journeys to Olde Englande', *Environment and Planning D: Society and Space* 12: 341–55.

—— (1996), 'Magic Kingdom or a Quixotic Quest for Authenticity?', *Annals of Tourism Research* 23: 415–31.

—— (1999), 'Knowing, Tourism and Practices of Vision', in Crouch, D. (ed.), *Leisure/Tourism Geographies: Practices and Geographical Knowledge*, London: Routledge.

Crang, M., Crang, P., and May, J. (1999), 'Introduction', in Crang, M., Crang, P., and May, J. (eds), *Virtual Geographies*, London: Routledge.

Cresswell, T. (2000), 'Mobility, Syphilis and Democracy: Pathologising the Mobile Body', in Wrigley, R., and Revill, G. (eds), *Pathologies of Travel*, Amsterdam: Clio Medica.

—— (2001), 'The Production of Mobilities', *New Formations* 43: 11–25.

Crouch, D. (1999), 'Introduction: Encounters in Leisure/Tourism', in Crouch, D. (ed.), *Leisure/Tourism Geographies: Practices and Geographical Knowledge*, London: Routledge.

—— (2002), 'Surrounded by Place – Embodied Encounters', in Coleman, S., and Crang, M. (eds), *Tourism: Between Place and Performance*, Oxford: Berghahn.

Csikszentmihalyi, M. (1990), *Flow: The Psychology of Optimal Experience*, New York: Harper & Row.

CTO (2001), *Tourism Strategy 2000–2010*, Nicosia: Cyprus Tourism Organization.

—— (2003), 'Cyprus: The Jewel of the Mediterranean', advertisement, London: Cyprus Tourism Organization.

Culin, S. (1899), 'Hawaiian Games', *American Anthropologist* 1: 201–47.

Culler, J. (1981), 'Semiotics of Tourism', *American Journal of Semiotics* 1: 127–40.

Curtin, P. (1990), *The Rise and Fall of the Plantation Complex: Essays in Atlantic History*, Cambridge: Cambridge University Press.

Cyprus Mail (2003), 'Brave Decisions Needed to Revive Tourism', *Cyprus Mail, Internet Edition*, 3 August; available online at *www.cyprus-mail.com/news/main.php?id=10651&archive=1*

Davis, A.(1991), *Dire Straits: The Dilemmas of a Fisher – The Case of Digby Neck and the Islands*, St John's: Institute of Social and Economic Research, Memorial University of Newfoundland.

Davis, Jeffrey (1994), 'Bombing Bikini Again, This Time With Money', *The New York Times Magazine*, 1 May.

De Carvalho, M.A.R. (1994), '"Governar por Retas": Engenheiros na Belle Èpoque Carioca', in *Quatro Vezes Cidade*, Rio de Janeiro: Sette Letras.

De Oliveira, A.C. (2001), 'Rocinha: a representação de um novo espaço social frente ao estado democrático de direito', M.A. thesis, Rio de Janeiro: Instituto Universitario de Pesquisas do Rio de Janeiro.

Degen, M. (2001), 'Regenerating Public Life? A Sensuous Comparison of Barcelona and Manchester', Ph.D. thesis, Lancaster University.

—— (2003), 'Fighting for the Global Catwalk: Formalising Public Life in Castlefield (Manchester) and Diluting Public Life in El Raval (Barcelona)', *International Journal of Urban and Regional Research* 27: 867–80.

Deleuze, G. (1992), 'Postscript on the Societies of Control', *October* 59: 3–7.

Delgado, M. (1992), 'La ciudad mentirosa', *El Basilico*, 2a Epoca, 12: 16–23.

DEOA (1995), *Dimension Endowment of Art*, Taipei: DEOA.

Desforges, L. (1997), '"Checking Out the Planet": Global Representations/Local Identities and Youth Travel', in Skelton, T., and Valentine, G. (eds), *Cool Places: Geographies of Youth Cultures*, London: Routledge.

—— (2000), 'Travelling the World: Identity and Travel Biography', *Annals of Tourism Research* 27: 926–45.

Dewailly, J-M. (1999), 'Sustainable Tourist Space: From Reality to Virtual Reality?', *Tourism Geographies* 1: 41–55.

Diken, B., and Lautsten, C.B. (2004), '7/11, 9/11, and Post-Politics', *Alternatives: A Journal of World Policy* 29 (forthcoming).

Discover Hong Kong (2003), 'Hong Kong, Live it, Love it', *http:webserv2.discoverhongkong.com/eng/* (accessed 6 February 2004).

Dodge, M., and Kitchin, R. (2001), *Mapping Cyberspace*, London: Routledge.

Dos Santos, M. (2001), 'The New Dynamic of Blockbuster Exhibitions: The Case of Brazilian Museums', *Bulletin of Latin American Research* 20: 29–45.

Duffy, R. (2000), 'Shadow Players: Ecotourism Development, Corruption and State Politics in Belize', *Third World Quarterly* 21: 549–65.

Duncan, J. (1999), 'Dis-orientation: On the Shock of the Familiar in a Far-Away Place', in Duncan, J., and Gregory, D. (eds), *Writes of Passage: Reading Travel Writing*, London and New York: Routledge.

Duval, D. (2002), 'The Return Visit–Return Migration Connection', in Hall, C.M., and Williams, A.M. (eds), *Tourism and Migration*, Dordrecht: Kluwer.

Edel, U. (director) (1981), *Christiane F: A True Story*, Second Sight Films.

Edensor, T. (1998), *Tourists at the Taj: Performance and Meaning at a Symbolic Site*, London: Routledge.

—— (2000), 'Staging Tourism: Tourists as Performers', *Annals of Tourism Research* 27: 322–44.

—— (2001a), 'Performing Tourism, Staging Tourism: (Re)Producing Tourist Space and Practice', *Tourist Studies* 1: 59–82.

—— (2001b), 'Walking in the British Countryside: Reflexivity, Embodied Practices and Ways to Escape', in Macnaghten, P., and Urry, J. (eds), *Bodies of Nature*, London: Sage.

—— (2004a), *Industrial Ruins: Uses, Order, Materiality and Memory*, Oxford: Berg.

—— (2004b), 'Sensing Tourist Spaces', in Minca, C., and Oakes, T. (eds), *Tourism and the Paradox of Modernity*, Minneapolis: University of Minnesota Press.

El Periodico (2002), 'El Maremagnum pierde 2 millones de visitantes por la conflictividad', February 2.

Elkins, T. H. (1988), *Berlin: The Spatial Structure of a Divided City*, London: Methuen.

Ellis, C. (1986), *Fisher Folk: Two Communities on Chesapeake Bay*, Lexington: University Press of Kentucky.

Elsrud, T. (1998), 'Time Creation in Travelling: The Taking and Making of Time among Women Backpackers', *Time and Society* 7: 309–34.

—— (2001), 'Risk Creation in Travelling: Backpacker Adventure Narration', *Annals of Tourism Research* 28: 597–617.

Enloe, C. (1989), *Bananas, Beaches and Bases: Making Feminist Sense of International Politics*, London: Pandora.

Fainstein, S., and Judd, D. (1999), 'Global Forces, Local Strategies and Urban Tourism', in Judd, D., and Fainstein, S. (eds), *The Tourist City*, London: Yale University Press.

Fang, J. (1998), 'Sanxingdui: The High Era of the Civilisation of Yangtze River' *Artists* June: 277.

Fang, S. (1993), 'Expecting the NPM to Get Out of the Isolated Palace', *Commonwealth Magazine*, Feb.: 54–58.

Farmer, P. (1992), *AIDS and Accusation: Haiti and the Geography of Blame*, Berkeley: University of California Press.

—— (1999), *Infections and Inequalities: The Modern Plagues*, Berkeley: University of California Press.

Farrell, S. (2002), 'The Taj Mahal: Pollution and Tourism', *TED Case Studies* 668 *www.american.edu/ted/taj.htm.*

Featherstone, M. (1991), *Consumer Culture and Postmodernism*, London: Sage.

Febvre, L.P.V. (1982), *The Problem of Unbelief in the Sixteenth Century*, translated by Beatrice Gottlieb, Cambridge, MA: Harvard University Press.

Feifer, M. (1985), *Going Places: The Ways of the Tourist from Imperial Rome to the Present Day*, London: Macmillan.

Fenby, J. (2002), 'The Day Jean-Marie Le Pen Opened a Pandora's Box of Bigotry', *Guardian*, 28 April; available online at *http://observer.guardian.co.uk/review/story/0,6903,706258,00.html*

Flores Galindo, A. (1987), *Buscando un Inca, Identidad y Utopia en Los Andes*, Cuzco: Instituto de Apoyo Agrario.

Foner, N. (1978), *Jamaica Farewell: Jamaican Migrants in London*, London: Routledge & Kegan Paul.

—— (1998), 'Towards a Comparative Perspective on Caribbean Migration', in Chamberlain, M. (ed.), *Caribbean Migration: Globalised Identities*, London: Routledge.

Forster, E.M. [1910] (1941), *Howard's End*, Harmondsworth: Penguin.

—— [1908] (1955), *A Room with a View*, Harmondsworth: Penguin.

Franklin, S., Lury, C., and Stacey, J. (2000), *Global Nature, Global Culture*, London: Routledge.

Freeman, C. (2000), *High Tech and High Heels in the Global Economy*, Durham, NC: Duke University Press.

Freeman, J. (2002), 'Democracy and Danger on the Beach: Class Relations in the Public Space of Rio de Janeiro', *Space and Culture* 5: 9–28.

Frith, S. (1997), 'The Suburban Sensibility in British Rock and Pop', in Silverstone, R. (ed.), *Visions of Suburbia*, London: Routledge.

Frow, J. (1997), *Time and Commodity Culture*, Oxford: Clarendon Press.

Fussell, P. (1980), *Abroad: British Literary Travelling between the Wars*, Oxford: Oxford University Press.

Galani-Moutafi, V. (2000), 'The Self and the Other: Traveller, Ethnographer, Tourist', *Annals of Tourism Research* 27: 203–24.

García, S. (1993), 'Barcelona und die Olympischen Spiele', in Hausserman, H., and Siebel, W. (eds), *Festivalisierung der Stadtpolitik*, Opladen: Westdeutscher Verlag.

—— (2003), 'The Case of Barcelona', in Salet, W., Thornley, A., and Kreukels, A. (eds), *Metropolitan Governance and Spatial Planning*, London: Spon Press.

García, S., and Claver, N. (2003), 'Governing Coalitions, Visitors, and the Changing City Centre', in Hoffman, L.M., Fainstein, S., and Judd, D. (eds), *Cities and Visitors*, London: Blackwell.

Garland, A. (1997), *The Beach*, Harmondsworth: Penguin.

Gdaniec, C. (2000), 'Cultural Industries, Information Technology and the Regeneration of Post-Industrial Urban Landscapes: Poblenou in Barcelona – a Virtual City?', *GeoJournal* 50: 379–87.

Gibson, H., and Yiannakis, A. (2002), 'Tourist Roles: Needs and the Lifecourse', *Annals of Tourism Research* 29: 358–83.

Gilroy, P. (1993), *The Black Atlantic: Modernity and Double Consciousness*, London and New York: Verso.

Glissant, E. (1981), *Le Discours Antillais*, Paris: de Seuil.

Goin, P. (1991), *Nuclear Landscapes*, Baltimore, MD: Johns Hopkins University Press.

Gottdiener, M. (1997), *The Theming of America: Dreams, Visions and Commercial Spaces*, Boulder, CO: Westview Press.

Gottlieb, A. (1982), 'Americans' Vacations', *Annals of Tourism Research* 9: 165–87.

Graham, B., Ashworth, G., and Tunbridge, J. (2000), *A Geography of Heritage: Power, Culture and Economy*, London: Arnold.

Graham, S., and Marvin, S. (2001), *Splintering Urbanism: Networked Infrastructures, Technological Mobilities and the Urban Condition*, London and New York: Routledge.

Greene, R. (2000), *Unrequited Conquest: Love and Empire in the Colonial Americas*, Chicago: University of Chicago Press.

Gregory, D. (1999), 'Scripting Egypt: Orientalism and the Cultures of Travel', in Duncan, J., and Gregory, D. (eds), *Writes of Passage: Reading Travel Writing*, London and New York: Routledge.

Gross, M. (1998), 'Anniversary Issue – Thomas Hoving: Reinventing the Museum', *New York Magazine*, 6 April.

Grove, R. (1995), *Green Imperialism: Colonial Expansion, Tropical Island Edens and the Origins of Environmentalism, 1600–1860*, Cambridge: Cambridge University Press.

Gunst, L. (1995), *Born fi' Dead: A Journey through the Jamaican Posse Underworld*, Edinburgh: Payback Press.

Halewood, C., and Hannam, K. (2001), 'Viking Heritage Tourism, Authenticity and Commodification', *Annals of Tourism Research* 28: 565–80.

Hall, D. (1992), 'Tourism Development in Cuba', in Harrison, D. (ed.), *Tourism and the Less Developed Countries*, Chichester: John Wiley.

Hall, P. (2000), 'Innovative Cities', in Hagbarth, I. (ed.), *Structural Change in Europe: Innovative Cities and Regions*, Bollschweil: Hagbarth.

Hall, S. (1990), 'Cultural Identity and Diaspora', in Rutherford, J. (ed.), *Identity: Community, Culture, Difference*, London: Lawrence & Wishart.

—— (1996), 'When Was the "Post-Colonial"? Thinking at the Limit', in Chambers, I., and Curti, L. (eds), *The Post-Colonial Question*, London: Routledge.

Hall, Sarah (2002), *Haweswater*, London: Faber & Faber.

Hall, T., and Hubbard, R. (1998), *The Entrepreneurial Cities*, Chichester: John Wiley.

Hall, C.M., Sharples, L., Mitchell, R., Macionis, N., Cambourne, B. (2003) *Food Tourism Around the World*, Amsterdam: Butterworth Heinemann

Hampton, M.P. (1998), 'Backpacker Tourism and Economic Development', *Annals of Tourism Research* 25: 639–60.

Hanna, S., and Del Casino Jr, V. (2003), 'Introduction: Tourism Spaces, Mapped Representations, and the Practices of Identity', in Hanna, S., and Del Casino Jr, V. (eds), *Mapping Tourism*, Minneapolis: University of Minnesota Press.

Hannigan, J. (1998), *Fantasy City: Pleasure and Profit in the Postmodern Metropolis*, London: Routledge.

Hansen, M. (1980), 'Traces of Transgression in Apocalypse Now', *Social Text* 3: 123–35.

Harding, L. (2000), 'Fury as India Cashes in on the Allure of the Taj Mahal', *The Observer*, 17 December; available online at
http://observer.guardian.co.uk/international/story/0,6903,412417,00.html

Harrogate Borough Council (2003), statistics, *www.harrogate.gov.uk/tourism/* (accessed 30 July 2003).

Harvey, D. (1985), 'The Geopolitics of Capitalism', in Gregory, D., and Urry, J. (eds), *Social Relations and Spatial Structures*, London: Macmillan.

—— (1989), *The Condition of Postmodernity*, Oxford: Blackwell.

Hattersley, R. (1976), *Goodbye to Yorkshire*, second edition, London: Pan.

Heelas, P. (1996), *The New Age Movement*, Oxford: Blackwell.

Heeren, S. von (2002), *La Remodelación de Ciutat Vella: Un Análisis Crítico del Modelo Barcelona*, Barcelona: Veïns en Defensa de la Barcelona Vella.

Heidegger, M. [1927] (1962), *Being and Time*, Oxford: Blackwell.

Heinich, N., and Pollak, M. (1996), 'From Museum Curator to Exhibition *Auteur*: Inventing a Singular Position', in Greenberg, R., Ferguson, B.W., and Nairne, S. (eds), *Thinking about Exhibitions*, London: Routledge.

Henderson, M. (2001), 'A Shifting Line Up: Men, Women, and *Tracks* Surfing Magazine', *Continuum: Journal of Media & Cultural Studies* 15: 318–32.

Hendry, J. (2000), *The Orient Strikes Back: A Global View of Cultural Display*, Oxford: Berg.

Henley, J. (2001), 'No-Go for Doctors in Paris Suburbs', *Guardian,* 24 April; available online at *www.guardian.co.uk/international/story/0,3604,477411,00.html*

Hermann, K., and Rieck, H. (1978), *Christiane F. Wir Kinder vom Bahnhof Zoo*, Düsseldorf: Stern.

Herriot, J. (1979), *James Herriot's Yorkshire,* London: Michael Joseph.

Hetherington, K. (1997), 'In Place of Geometry: The Materiality of Place', in Hetherington, K., and Munro, R. (eds), *Ideas of Difference*, Oxford: Blackwell.

—— (1999), 'From Blindness to Blindness: Museums, Heterogeneity and the Subject', in Law, J., and Hassard, J. (eds), *Actor Network Theory and After*, London: Blackwell.

—— (2000), *New Age Travellers: Vanloads of Uproarious Humanity*, London: Continuum.

—— (2002), 'The Unsightly: Touching the Parthenon Frieze', *Theory, Culture and Society* 19: 187–206.

Hoban, G. *et al.* (1994), *Black Wind White Land: Living with Chernobyl*, Dublin: Dream Chaser Productions.

Hoggett, P., and Bishop, J. (1986), *Organizing Around Enthusiasms*, London: Comedia.

Holmes, D. (2001), 'Virtual Globalization – an Introduction', in Holmes, D. (ed.), *Virtual Globalization: Virtual Spaces/Tourist Spaces*, London: Routledge.

Hong Kong Government (2003), 'Hong Kong Looking Ahead', *www.hklookingahead.gov.hk/home/* (accessed 21 September 2003).

Hooks, B. (1992), 'Eating the Other', in *Black Looks: Race and Representation*, London: Turnaround.

Hooper-Greenhill, E. (1992), *Museums and the Shaping of Knowledge*, London: Routledge.

Horwitz, J., and Singley, P. (eds) (2004), *Eating Architecture*, Cambridge, MA: M.I.T. Press.

Hoving, T. (1993), *Making the Mummies Dance: Inside the Metropolitan Museum of Art*, New York: Simon & Schuster.

Howe, C. (1998), 'The Taiwan Economy: The Transition to Maturity and the Political Economy of its Changing International Status', in Shambaugh, D. (ed.), *Contemporary Taiwan*, Oxford: Oxford University Press.

Huang, C. *et al.* (1996), 'Trends in Outbound Tourism from Taiwan', *Tourism Management* 17: 223–28.

Huang, K. (1997), *Museum Marketing Strategies*, Taipei: Artist Press.

Hughes, R. (1968), *Hong Kong: Borrowed Place, Borrowed Time*, London: André Deutsch.

Huizinga, J. (1950), *Homo Ludens: A Study of the Play Element in Culture*, Boston: Beacon.

Ingold, T. (2000), *The Perception the Environment: Essays on Livelihood, Dwelling and Skill*, London: Routledge.

INM Asia Guide (2003), *Hong Kong Travel Guide*, *www.inm-asiaguides.com/hongkong/ehkgkow.htm* (accessed 6 February 2004).

Ioannides, D. (1992), 'Tourism Development Agents: the Cypriot Resort Cycle', *Annals of Tourism Research* 19: 711–31.

Ioannides, D., Apostolopoulos, Y., and Sonmez, S. (eds) (2001), *Mediterranean Islands and Sustainable Tourism Development: Practices, Management and Policies*, London: Continuum.

Ishiwata, E. (2002), 'Local Motions: Surfing and the Politics of Wave Sliding', *Cultural Values* 6: 257–72.

Jenkins, A. (2001), 'Lust for Life', *Guardian*, Travel Section, 17 February 2001: 6; available online at *http://travel.guardianunlimited.co.uk/saturdaysection/story/0,8922,438914,00.html*

Jenkinson, D.J. (1998), *Then and Now: Harrogate*, Stroud: Tempus.

Jessop, B., and Sum, N-L. (2000), 'An Entrepreneurial City in Action: Hong Kong's Emerging Strategies in and for (Inter)Urban Competition', *Urban Studies* 37: 287–314.

Joyce, E. (1998), 'Cocaine Trafficking and British Foreign Policy', in Joyce, E., and Carlos, M. (eds), *Latin America and the Multinational Drug Trade*, Basingstoke: Macmillan.

Judd, D., and Fainstein, S. (1999), *The Tourist City*, New Haven, CT: Yale University Press.

Julier, G. (1996), 'Barcelona Design, Catalonia's Political Economy, and the New Spain, 1980–1986', *Journal of Design History* 9: 117–27.

Kampion, D., and Brown B. (2003), *A History of Surf Culture*, Cologne: Taschen.

Kaplan, C. (2003), 'Transporting the Subject: Technologies of Mobility and Location in an Era of Globalization', in Ahmed, S. *et al.* (eds), *Uprootings/Regroundings: Questions of Home and Migration*, Oxford: Berg.

Kazim, A. (2000), *The United Arab Emirates A.D. 600 to the Present: A Socio-Discursive Transformation in the Arabian Gulf,* Dubai: Gulf Book Centre.

Kearns, G., and Philo, C. (eds) (1993), *Selling Places*, Oxford: Pergamon.

Kempadoo, K. (ed.) (1999), *Sun, Sex, and Gold: Tourism and Sex Work in the Caribbean*, Lanham, MD: Rowman & Littlefield.

Kennaway, G. (2000), 'Road to a New Eden (with Plenty of Temptations along the Way)', *High Life* March: 114.

King, R. (1993), 'The Geographical Fascination of Islands', in Lockhart, D., Drakakis-Smith, D., and Schembri, J. (eds), *The Development Process in Small Island States*, London: Routledge.

Klak, T., and Myers, G. (eds) (1998), *Globalization and Neoliberalism: The Caribbean Context*, Lanham, MD: Rowman & Littlefield.

Klein, N. (2000), *No Logo*, London: Flamingo.

Knight, A., and Nakano, Y. (1999), *Reporting on Hong Kong: Foreign Media and the Handover*, New York: St Martin's Press.

Kolland, F., Milian, J., Scheibenstock, M., and Schönbauer, H. (2000), 'The Short Waves of the Product-Cycle of Tourism: Cuba's Tourism between De-coupling and Integration into the World Market', in Robinson, M. *et al.* (eds), *Management, Marketing and the Political Economy of Travel and Tourism*, Sunderland: Business Education Publishers.

Koolhaas, R., and Mau, B. (1997), *S, M, L, XL*, Cologne: Taschen.

Kotler, P. *et al.* (1993), *Marketing Places*, New York: Macmillan.

Lane, J. (2000), *Pierre Bourdieu*, London: Pluto Press.

Larsen, J. (2003), 'Performing Tourist Photography', Ph.D. thesis, Roskilde: Roskilde University.

Lash, S. (2001), 'Technological Forms of Life', *Theory, Culture and Society* 18: 105–20.

—— (2002), *Critique of Information*, London: Sage.

Lash, S., and Urry, J. (1994), *Economies of Signs and Space*, London: Sage.

Laurence, W.L. (1947), *Dawn Over Zero: The Story of the Atomic Bomb*, London: Museum Press.

Le Corbusier (1998), *Rio de Janeiro: 1929, 1936*, Rio de Janeiro: Centro de Arquitetura e Urbanismo do Rio de Janeiro, Prefeitura da Cidade do Rio de Janeiro.

Lee, L. (1972), *Travel Diary: Accompanying the Touring Exhibition*, Taipei: Taiwan Shang-Wu Press.

Lee, T.H., and Crompton, J. (1992), 'Measuring Novelty-Seeking in Tourism', *Annals of Tourism Research* 19: 732–51.

Leith, W. (2003), 'Give Me an Eau', *Observer Food Monthly*, 10 August: 16–22; available online at *http://observer.guardian.co.uk/foodmonthly/story/0,9950,1013279,00.html*

Lennon, J., and Foley, M. (2000), *Dark Tourism: The Attraction of Death and Disaster*, London: Continuum.

Lett, J. (1983), 'Ludic and Liminoid Aspects of Charter Yacht Tourism in the Caribbean', *Annals of Tourism Research* 10: 35–56.

Lewis, N. (2001), 'The Climbing Body, Nature and the Experience of Modernity', in Macnaghten, P., and Urry, J. (eds), *Bodies of Nature*, London: Sage.

Light, D. (2000), 'Gazing on Communism: Heritage Tourism and Post-Communist Identities in Germany, Hungary and Romania', *Tourism Geographies* 2: 157–76.

Lo, K-C. (2001), 'Giant Panda and Mickey Mouse: Transnational Objects of Fantasy in Post-1997 Hong Kong', *www.international.ucla.edu/cira/lo_paper.htm* (accessed 6 February 2004).

Loader, Jayne, Kevin Rafferty, and Pierce Rafferty. (1982). *The Atomic Cafe*, New York: The Archives Project, Inc.

Lockhart, D. (1997), 'Islands and Tourism: an Overview', in Lockhart, D., and Drakakis-Smith, D. (eds), *Island Tourism: Trends and Prospects*, London: Pinter.

Lodge, D. (1991), *Paradise News*, London: Secker & Warburg.

Löfgren, O. (1999), *On Holiday: A History of Vacationing*, Berkeley: University of California Press.

—— (2004), 'Narrating the Tourist Experience,' in Gmelch, S.B. (ed.), *Tourists and Tourism: A Reader*, Long Grove, IL: Waveland Press.

Logan, D.F. (1983), *The Vikings in History*, London: Routledge.

Lownhaupt-Tsing, A. (1994), 'From the Margins', *Cultural Anthropology* 9: 279–97.

Lübbren, N. (2001), *Rural Artists' Colonies in Europe 1870–1910*, Manchester: Manchester University Press.

Lury, C. (1997), 'The Objects of Travel', in Rojek, C., and Urry, J. (eds), *Touring Cultures: Transformations of Travel and Theory*, London: Routledge.

—— (1998), *Prosthetic Culture: Photography, Memory and Identity*, London: Routledge.

Lyons, P. (1997), 'Pacific Scholarship, Literary Criticism, and Touristic Desire', *Boundary* 24: 47–78.

MacCannell, D. (1973), 'Staged Authenticity: Arrangements of Social Space in Tourist Settings', *American Journal of Sociology* 79: 589–603.

—— (1976), *The Tourist: A New Theory of the Leisure Class*, Berkeley: University of California Press.

—— (1989), *The Tourist: A New Theory of the Leisure Class*, second edition, New York: Schoken Books.

—— (1992), *Empty Meeting Grounds*, London: Routledge.

Macnaghten, P., and Urry, J. (1998), *Contested Natures*, London: Sage.

Macnaghten, P., and Urry, J. (eds) (2001), *Bodies of Nature*, London: Sage.

Maingot, A. (1993), 'The Offshore Caribbean', in Payne, A., and Sutton, P. (eds), *Modern Caribbean Politics*, Baltimore: Johns Hopkins University Press.

Malbon, B. (1998), 'The Club: Clubbing, Consumption, Identity and the Spatial Practices of Every-Night Life', in Skelton, T., and Valentine, G. (eds), *Cool Places: Geographies of Youth Cultures*, London: Routledge.

Mann, T. [1912] (1955), *Death in Venice*, Harmondsworth: Penguin.

Markus, G. (2001), 'Walter Benjamin, or: The Commodity as Phantasmagoria', *Theory, Culture and Society* 83: 3–42.

Markwell, K.W. (1997), 'Dimensions of Photography in a Nature-Based Tour', *Annals of Tourism Research* 24: 131–55.

Marshall, T. (2000), 'Urban Planning and Governance: Is There a Barcelona Model?', *International Planning Studies* 5: 299–319.

Martin, A. (2000), 'Beach', in Coates, S., and Stetter, A. (eds), *Impossible Worlds*, Boston: Birkhäuser.

Maspero, F. (1994), *Roissy Express: A Journey Through the Paris Suburbs*, London: Verso.

Massey, D. (1993), 'Power-Geometry and a Progressive Sense of Place', in Bird, J., Curtis, B., Putnam, T., Robertson, G., and Tickner, L. (eds), *Mapping the Futures: Local Cultures, Global Change*, London: Routledge.

—— (1994), *Space, Place and Gender*, Cambridge: Polity Press.

Matthews, D.R. (1993), *Controlling Common Property: Regulating Canada's East Coast Fishery*, Toronto: University of Toronto Press.

Maurer, B. (1995), 'Complex Subjects: Offshore Finance, Complexity Theory, and the Dispersion of the Modern', *Socialist Review* 25: 113–45.

—— (1997a), 'Creolization Redux: The Plural Society Thesis and Offshore Financial Services in the British Caribbean', *New West Indian Guide* 71: 249–64.

—— (1997b), *Recharting the Caribbean: Land, Law and Citizenship in the British Virgin Islands*, Ann Arbor: University of Michigan Press.

—— (2002), 'A Fish Story: Rethinking Globalization on Virgin Gorda, British Virgin Islands', in Inda, J.X., and Rosaldo, R. (eds), *The Anthropology of Globalization: A Reader*, Oxford: Blackwell.

Maxwell Stamp Plc (1991), *Belize: Tourism Planning Stage 1*, prepared for the (UK) Overseas Development Administration on behalf of the Government of Belize.

McCalla, W. (1995), *Guide for Developers*, Belmopan: Ministry of Tourism and the Environment, Department of the Environment, Belize.

McMinn, S., and Cater, E. (1998), 'Tourist Typology: Observations from Belize', *Annals of Tourism Research* 25: 675–99.

McNeill, D. (1999), *Urban Change and the European Left*, London: Routledge.

—— (2003), 'Mapping the European Left: The Barcelona Experience', *Antipode* 35: 74–94.

Meethan, K. (2001), *Tourism in Global Society: Place, Culture, Consumption*, New York: Palgrave.

Mendieta, E. (2001), 'Invisible Cities: a Phenomenology of Globalization from Below', *City* 5: 7–26.

Michael Hall, C., Sharples, L., Mitchell, R., Macionis, N., and Cambourne, B. (eds) (2003), *Food Tourism Around the World*, Oxford: Butterworth–Heinemann.

Milne, S. (1992), 'Tourism and Development in South Pacific Microstates', *Annals of Tourism Research* 19: 191–212.

—— (1997), 'Tourism, Dependency and South Pacific Microstates: Beyond the Vicious Circle?', in Lockhart, D., and Drakakis-Smith, D. (eds), *Island Tourism: Trends and Prospects*, London: Pinter.

Milton, K. (1993), 'Land or Landscape: Rural Planning Policy and the Symbolic Construction of the Countryside', in Murray, M., and Greer, J. (eds), *Rural Development in Ireland*, Aldershot: Avebury.

Minca, C., and Oakes, T. (2004a), 'Introduction: Travelling Paradoxes', in Minca, C., and Oakes, T. (eds), *Tourism and the Paradox of Modernity*, Minneapolis: University of Minnesota Press.

—— (eds) (2004b), *Tourism and the Paradox of Modernity*, Minneapolis: University of Minnesota Press.

Ministry of the Interior, Taiwan (1994), *Statistics of the Ministry of the Interior*, Taipei: Ministry of the Interior.

Misrach, R. (1990), *Bravo 20: The Bombing of the American West*, Baltimore, MD: Johns Hopkins University Press.

Mitra, A., and Cohen, E. (1999), 'Analyzing the Web: Directions and Challenges', in Jones, S.G. (ed.), *Doing Internet Research*, Thousand Oaks, CA: Sage.

Molotch, H. (2002), 'Place in Product', *International Journal of Urban and Regional Research* 26: 665–88.

Moreno, E., and Montalbán, V. (1991), *Barcelona, cap a on vas?: diàlegs per a una altra Barcelona*, Barcelona: Llibres de l'Index.

Morris, J. (1997), *Hong Kong: Epilogue to an Empire*, Atlanta: Vintage.

Mowforth, M., and Munt, I. (1998), *Tourism and Sustainability: New Tourism in the Third World*, London: Routledge.

—— (2003), *Tourism and Sustainability: Development and Tourism in the Third World* (second edition of 1998 title), London: Routledge.

Mullings, B. (1999), 'Globalization, Tourism and the International Sex Trade', in Kempadoo, K. (ed), *Sun, Sex, and Gold: Tourism and Sex Work in the Caribbean*, Lanham, MD: Rowman & Littlefield.

Munt, I. (1994a), 'The "Other" Postmodern Tourism: Culture, Travel and the New Middle Classes', *Theory, Culture and Society* 11: 101–23

—— (1994b), 'Ecotourism or Egotourism?', *Race and Class* 36: 49–60.

Murphy, B. (2000), *Home Truths: A Jaunt Round the Decaying Heart of England*, Edinburgh: Mainstream.

Murphy, L. (2001), 'Exploring Social Interactions of Backpackers', *Annals of Tourism Research* 28: 50–67.

Nash, C. (2003), ' "They're Family!": Cultural Geographies of Relatedness in Popular Genealogy', in Ahmed, S., Fortier, A-M., Castaneda, C., and Sheller, M. (eds), *Uprootings/Regroundings: Questions of Home and Migration*, Oxford: Berg.

Neesam, M. (1983), *Harrogate in Old Picture Postcards*, Zaltbommel: European Library.

—— (1999), *Bygone Harrogate*, Derby: Breedon Books.

—— (2001), *Harrogate: A History of the English Spa from Earliest Times to the Present*, Stroud: Tempus.

Negt, O., and Kluge, A. (1987), *Geschichte und Eigensinn* [History and Obstinacy], Frankfurt am Main: Zweitausendundeins.

—— (1993), *Public Sphere and Experience*, Minneapolis: University of Minnesota Press.

Nevarez, L. (2002), *New Money, Nice Town: How Capital Works in the New Urban Economy*, New York: Routledge.

Newmann, M. (2002), 'Making the Scene: The Poetics and Performances of Displacement at the Grand Canyon', in Coleman, S., and Crang, M. (eds), *Tourism: Between Place and Performance*, New York: Berghahn.

Noble Caledonian Ltd (2002), 'West Indies: Hidden Treasures: 14-Night Cruise on the *Levant*', 8–23 February 2002, advertised in the *Financial Times*.

Nunes, M. (1997), 'What Space is Cyberspace?', in Holmes, D. (ed.), *Virtual Politics: Identity and Community in Cyberspace*, London: Sage.

O'Rourke, P.J. (1989), *Holidays in Hell*, London: Picador.

Oak, P. (1994), 'The Tajmahal is Tejomahalay, a Hindu Temple', at *http://www.hindunet.org/hindu_history/modern/taj_oak.html* (accessed 6 February 2004).

Olsen, K. (2002), 'Authenticity as a Concept in Tourism Research', *Tourist Studies* 2: 159–81.

Olwig, K.F. (1993), *Global Culture, Island Identity*, Philadelphia, PA: Harwood.

Osborne, P. (2000), *Travelling Light: Photography, Travel and Visual Culture*, Manchester: Manchester University Press.

Ostwald, M.J. (2000), 'Virtual Urban Futures', in Bell, D., and Kennedy, B.M. (eds), *The Cybercultures Reader*, London: Routledge.

Ousby, I. (1990), *The Englishman's England*, Cambridge: Cambridge University Press.

Pang, L-K. (2002), 'Sightseeing an (Inter)national City: Hong Kong Tourism and the Society of the Spectacle', *www.international.ucla.edu/cira/pang_paper.htm* (accessed 6 February 2004).

Parr, M. (1999), *Boring Postcards*, London: Phaidon.

—— (2000), *Boring Postcards USA*, London: Phaidon.

—— (2001), *Langweilige Postkarten*, Berlin: Phaidon.

Pattullo, P. (1996), *Last Resorts: The Cost of Tourism in the Caribbean*, London: Cassell.

—— (2004), 'Sailing into the Sunset: The Cruise-ship Industry', in Gmelch, S.B. (ed.), *Tourists and Tourism: A Reader*, Long Grove, IL: Waveland Press.

Pemble, J. (1987), *The Mediterranean Passion*, Oxford: Clarendon Press.

Philpott, S.B. (1973), *West Indian Migration: The Montserrat Case*, London: Athlone.

Phipps, P. (1999), 'Tourists, Terrorists, Death and Value', in Kaur, R., and Hutnyk, J. (eds), *Travel Worlds*, London: Zed Books.

Pietrasik, A. (2001), 'Pleasure Island', *Guardian*, Travel Section, 11 August: 12–13; available online at *http://travel.guardianunlimited.co.uk/saturdaysection/story/0,8922,534895,00.html*

PIO (2001), *About Cyprus*, Nicosia: Press and Information Office.

Pi-Sunyer, O. (1996), 'Tourism in Catalonia', in Barke, M., Towner, J., and Newton, M.T. (eds), *Tourism in Spain*, Wallingford, Oxfordshire: C.A.B. International.

Poe, Edgar Allan [1827] (1993), 'A Dream within a Dream', in *Complete Poems and Selected Essays*, edited by R. Gray, New York: Everyman.

Poole, D. (1998), 'Landscape and the Imperial Subject: U.S. Images of the Andes, 1859–1930', in Joseph, G., Legrand, C., and Salvatore, R. (eds), *Close Encounters of Empire*, Durham, NC: Duke University Press.

Pratt, M.L. (1992), *Imperial Eyes: Travel Writing and Transculturation*, London: Routledge.

Preston-Whyte, R. (2002), 'Constructions of Surfing Space at Durban, South Africa', *Tourism Geographies* 4: 307–28.

Raz, A. (1999), *Riding the Black Ship: Japan and Tokyo Disneyland*, Cambridge, MA: Harvard University Press.

Redfoot, D.L. (1984), 'Touristic Authenticity, Touristic Angst and Modern Reality', *Qualitative Sociology* 7: 291–309.

Relph, E. (1976), *Place and Placelessness*, London: Pion.

Rey, H. (1996), *La Peur des Banlieues*, Paris: Presses des Sciences Politiques.

Rheingold, H. (1994), *The Virtual Community*, London: Secker & Warburg.

Richards, J. (2001), *Blood of the Vikings,* London: Hodder & Stoughton.

Ring, J. (2000), *How the English Made the Alps*, London: John Murray.

Ringer, G. (ed.) (1998), *Destinations*, London: Routledge.

Ritzer, G. (1996), *The McDonaldization of Society*, California: Pine Forge Press.

Ritzer, G., and Liska, A. (1997), 'McDisneyization and Post-Tourism', in Rojek, C., and Urry, J. (eds), *Touring Culture: Transformation of Travel and Theory*, London: Routledge.

ROC Tourism Bureau (1996), *Annual Report on Tourism Statistics*, Taipei: Ministry of Transportation and Communication.

Roche, A. (1996), *Children of Chernobyl: The Human Cost of the World's Worst Nuclear Disaster*, London: HarperCollins.

Roche, M. (2000), *Mega-Events and Modernity: Olympics and Expos in the Growth of Global Culture*, London: Routledge.

Rojek, C. (1993), *Ways of Escape*, Basingstoke: Macmillan.

—— (1995), *Decentring Leisure, Rethinking Leisure Theory*, London: Sage.

—— (1997), 'Indexing, Dragging and the Social Construction of Tourist Sights', in Rojek, C., and Urry, J. (eds), *Touring Cultures: Transformations of Travel and Theory*, London: Routledge.

—— (1998) 'Cybertourism and the Phantasmagoria of Space', in Ringer, G. (ed.), *Destinations*, London: Routledge.

Rojek, C., and Urry, J. (1997), 'Transformations of Travel and Theory', in Rojek, C., and Urry, J. (eds), *Touring Cultures: Transformations of Travel and Theory*, London: Routledge.

Rothman, H. (2002), *Neon Metropolis: How Las Vegas Started the Twenty-First Century*, New York: Routledge.

Rundle, M.L. (2001), 'Tourism, Social Change and *Jineterismo* in Contemporary Cuba', paper presented at the 25th Annual Conference of the Society for Caribbean Studies, University of Nottingham, July.

Rutsky, R.L. (1999), 'Surfing the Other: Ideology of the Beach', *Film Quarterly* 52: 12–23.

Rydell, R. (1984), *All the World's a Fair: Visions of Empire at American International Expositions, 1876–1916*, Chicago: Chicago University Press.

Sánchez Taylor, J. (1999), 'Tourism and "Embodied" Commodities: Sex Tourism in the Caribbean', in Clift, S., and Carter, S. (eds), *Tourism and Sex: Culture, Commerce and Coercion*, London and New York: Pinter.

Sassen, S. (2001), *The Global City*, second edition, Princeton, NJ: Princeton University Press.

Sassen, S., and Roost, F. (1999), 'The City: Strategic Site for the Global Entertainment Industry', in Judd, D., and Fainstein S. (eds), *The Tourist City*, London: Yale University Press.

Schroeder, J. (2002), *Visual Consumption*, London: Routledge.

Schultz, P. (2003), *1000 Places to See Before You Die*, New York: Workman.

Schwartz, R. (1997), *Pleasure Island: Tourism and Temptation in Cuba*, Lincoln: University of Nebraska Press.

Scott, A. (2001), 'Capitalism, Cities, and the Production of Symbolic Forms', *Transactions Institute of British Geographers* 26: 11–23.

Scott, D., and Willits, F.K. (1994), 'Environmental Attitudes and Behaviour: A Pennsylvania Survey', *Environment and Behaviour* 26: 239–60.

Seekings, J. (1997), *Cyprus* (International Tourism Reports 4), London: Economic Intelligence Unit.

Sellars, A. (1996), 'The Influence of Dance Music on the UK Youth Tourism Market', *Tourism Management* 19: 611–15.

Sennett, R. (1991), *The Conscience of the Eye,* London: Faber & Faber.

—— (1994), *Flesh and Stone*, London: Faber & Faber.

Sharpley, R. (1998a), *Island Tourism Development: The Case of Cyprus*, Sunderland: Business Education Publishers.

—— (1998b), *Sustainable Tourism Development: A Theoretical and Empirical Analysis*, Ph.D. thesis: Lancaster University.

—— (2001a), 'Tourism in Cyprus: Challenges and Opportunities', *Tourism Geographies* 3: 64–85.

—— (2001b), 'Sustainability and the Political Economy of Tourism in Cyprus', *Tourism* 49: 241–54.

—— (2002), 'Rural Tourism and the Challenge of Diversification: The Case of Cyprus', *Tourism Management* 23: 233–44.

—— (2003a), 'Tourism, Modernisation and Development on the Island of Cyprus: Challenges and Policy Responses', *Journal of Sustainable Tourism* 11: 246–65.

—— (2003b), 'Travels to the Edge of Darkness: Towards a Typology of "Dark Tourism"', paper given to the *Taking Tourism to the Limits* Conference, Waikato University, Hamilton, 9–11 December.

Sharpley, R., and Telfer, D. (eds) (2002), *Tourism and Development: Concepts and Issues*, Clevedon: Channel View Publications.

Sheller, M. (2000), *Democracy after Slavery: Black Publics and Peasant Radicalism in Haiti and Jamaica*, London: Macmillan.

—— (2003a), *Consuming the Caribbean: From Arawaks to Zombies*, London and New York: Routledge.

—— (2003b), 'Creolization in Global Culture', in Ahmed, S., Fortier, A.M., Castaneda, C., and Sheller, M. (eds), *Uprootings/Regroundings: Questions of Home and Migration,* Oxford: Berg.

—— (2004), 'Natural Hedonism', in Duval, D. (ed.), *Tourism in the Caribbean*, London and New York: Routledge.

Sheller, M., and Urry, J. (2000), 'The City and the Car', *International Journal of Urban and Regional Research* 24: 737–57.

—— (2003), 'Mobile Transformations of "Public" and "Private" Life', *Theory, Culture and Society* 20: 107–25.

Shields, R. (1991), *Places on the Margin: Alternative Geographies of Modernity*, London: Routledge.

—— (1996), 'Introduction: Virtual Spaces, Real Histories and Living Bodies', in Shields, R. (ed.), *Cultures of Internet*, London: Sage.

—— (1999), *Lefebvre: Love and Struggle – Spatial Dialectics*, London: Routledge.

—— (2003), *The Virtual*, London: Routledge.

Short, J., and Kim, Y-H. (1999), *Globalization and the City*, New York: Longman.

Silverstone, R. (ed.) (1997), *Visions of Suburbia*, London: Routledge.

Simmel, G. [1903] (1950), 'The Metropolis and Mental Life', in Wolff, K. (ed.), *The Sociology of Georg Simmel*, New York: Free Press.

—— [1896] (1997), 'The Berlin Trade Exhibition', in Frisby, D., and Featherstone, M. (eds), *Simmel on Culture*, London: Sage.

Singh, H. (2003), 'Analysis: Halting Work around the Taj Mahal Praised', *The Washington Times*, *www.washtimes.com/upi-breaking*.

Skelton, T. and Valentine, G. (eds) (1998), *Cool Places: Geographies of Youth Cultures*, London: Routledge.

Skinner, E. (1998), 'The Caribbean Data Processors', in Sussman, G., and Lent, J. (eds), *Global Productions: Labor in the Making of the 'Information Society'*, Cresskill, NJ: Hampton Press.

Smith, M. (2003), *Issues in Cultural Tourism Studies*, London: Routledge.

Smith, V. (ed.) (1989), *Hosts and Guests: The Anthropology of Tourism*, Philadelphia: University of Pennsylvania Press.

Smyth, H. (1994), *Marketing the City*, London: Spon Press.

Squire, S. (1998), 'Rewriting Languages of Geography and Tourism: Cultural Discourses of Destinations, Gender and Tourism History in the Canadian Rockies', in Ringer, G. (ed.), *Destinations: Cultural Landscapes of Tourism*, London: Routledge.

Strachan, I. (2002), *Paradise and Plantation: Tourism and Culture in the Anglophone Caribbean*, Charlottesville: University of Virginia Press.

Stranger, M. (1999), 'The Aesthetics of Risk: A Study of Surfing', *International Review for the Sociology of Sport* 34: 265–76.

Stratton, J. (2000), 'Cyberspace and the Globalization of Culture', in Bell, D., and Kennedy, B.M. (eds), *The Cybercultures Reader*, London: Routledge.

Subiros, P. (1999), *Estratègies culturals i renovació urbana*, Barcelona: Aula Barcelona.

Sum, N-L. (1999), 'Rethinking Globalization: Rearticulating the Spatial Scales and Temporal Horizons of Trans-border Space', in Olds, K. *et al.* (eds), *Globalization in the Asia Pacific*, London: Routledge.

—— (2002), 'Re-articulation of Spatial Scales and Temporal Horizons of a Cross-Border Mode of Growth: the (Re-)Making of "Greater China"', in Perkmann, M., and Sum, N-L. (eds), *Globalization, Regionalization and Cross-Border Regions*, London: Palgrave.

—— (2003), 'Informational Capitalism and US Economic Hegemony: Resistance and Adaptations in East Asia', *Critical Asian Studies* 35: 373–98.

Supurrier, J. (2002), 'A Generation of Gidgets: Female Surfers are Back in the Lineup, in Droves', *Atlantic Monthly* April: 109–11.

Switzer, T. (2002), 'Hungarian Spas', in Anderson, S., and Tabb, B. (eds), *Water, Leisure and Culture*, Oxford: Berg.

Syal, R. (2002), 'Women's Face-Pack Recipe Restores the Taj Mahal to its Former Glory', *The Sydney Morning Herald*, 2 December; available online at
www.smh.com.au/articles/2002/12/01/1038712831021.html

Szerszynski, B., and Urry, J. (2002), 'Cultures of Cosmopolitanism', *Sociological Review* 50: 461–81.

Taylor, R. (2002), *House Inside the Waves: Domesticity, Art and the Surfing Life*, Vancouver: Beach Holme Publishing.

Tello i Robira, R. (1993), 'Barcelona Post-Olimpica: de Ciudad Industrial a Ciudad de Consumo', *Estudios Geograficos* 212: 507–19.

Tester, K. (ed.) (1994), *The Flâneur*, London: Routledge.

Thomas-Hope, E. (1992), *Explanation in Caribbean Migration: Perception and the Image*, London: Macmillan.

—— (1995), 'Island Systems and the Paradox of Freedom: Migration in the Post-emancipation Leeward Islands', in Olwig, K.F. (ed.), *Small Islands, Large Questions*, London: Frank Cass.

Thompson, G. (2002), *Hazard Posed by UK Nuclear Facilities: Whose Position is Correct?*, Cambridge, MA: Institute for Resource and Security Studies.

—— (2003), *Robust Storage of Spent Nuclear Fuel: A Neglected Issue of Homeland Security*, Cambridge, MA: Institute for Resource and Security Studies.

Till, K. (2003), 'Construction Sites and Showcases: Mapping "The New Berlin" through Tourism Practices', in Hanna, S., and Del Casino Jr, V. (eds), *Mapping Tourism*, Minneapolis: University of Minnesota Press.

Time Out Guide (2002), *Barcelona*, London: Penguin.

Titley, G. (2000), 'Global Theory and Touristic Encounters', *Irish Communications Review* 8: 79–87.

Tolkien, J.R.R. (1997), *The Monsters and the Critics, and Other Essays*, London: HarperCollins.

Tung, C-H. (1998), 'Policy Address 1998, From Adversity to Opportunity', 7 October 1998; available online at *www.policyaddress.gov.hk/pa98/english/econ2.htm*.

—— (2001), 'Policy Address 2001, Building On Our Strengths, Investing in Our Future', 10 October 2001; available online at *www.policyaddress.gov.hk/pa01/e79.htm* (accessed on 12 October 2003).

Turkle, S. (1995), *Life on the Screen: Identity in the Age of the Internet*, New York: Simon & Schuster, and London: Weidenfeld & Nicolson.

Turner, V. (1969), *The Ritual Process*, London: Penguin.

Twain, Mark [1872] (1972), *Roughing It*, Berkeley: University of California Press.

Urry, J. (1990), *The Tourist Gaze: Leisure and Travel in Contemporary Societies*, London: Sage.

—— (1992), 'The Tourist Gaze and the Environment', *Theory, Culture and Society* 9: 3–24.

—— (1994), 'Cultural Change and Contemporary Tourism', *Leisure Studies* 13: 233–38.

—— (1995), *Consuming Places*, London: Routledge.

—— (2000), *Sociology Beyond Societies: Mobilities for the Twenty-First Century*, London: Routledge.

—— (2002a), *The Tourist Gaze*, second edition, London: Sage.

—— (2002b), 'Mobility and Proximity', departmental paper, Lancaster University: Sociology Department.

—— (2003a), *Global Complexity*, Cambridge: Polity Press.

—— (2003b), 'Social Networks, Travel and Talk', *British Journal of Sociology* 54: 155–75.

Ventura, Z. (1994), *A cidade partida*, São Paulo: Companhia das Letras.

Visconti, L. (director) (1971), *Death in Venice*, Warner Brothers Film.

Wallis, B. (1994), 'Selling Nations: International Exhibitions and Cultural Diplomacy', in Sherman, D.J., and Rogoff, I. (eds), *Museum/Culture*, London: Routledge.

Walsh, K. (1992), *The Representation of the Past: Museums and Heritage in the Post-Modern World*, London: Routledge.

Walt Disney Company (Asia Pacific) (2003), 'Disney and Hong Kong Government Break Ground on First Theme Park in China', *www.thetimesharebeat.com/2003/htl/jan/0113-03h.htm* (accessed 6 February 2004).

Wang, N. (1999), 'Rethinking Authenticity in Tourism Experience', *Annals of Tourism Research* 26: 349–70.

—— (2000), *Tourism and Modernity: A Sociological Analysis*, Oxford: Pergamon.

Wang, Y., Yu, Q., and Fesenmaier, D.R. (2002), 'Defining the Virtual Tourist Community: Implications for Tourism Marketing,' *Tourism Management* 23/4: 407–17.

Ward, N. (1996), 'Surfers, Sewage and the New Politics of Pollution', *Area* 28: 331–38.

Wardle, H. (1999), 'Jamaican Adventures: Simmel, Subjectivity and Extraterritoriality in the Caribbean', *Journal of the Royal Anthropological Institute* 5: 523–39.

Watson, P. (2003),'Targeting Tourists, Not Terrorists: Why Airport Security is a Charade', 8 January; available online at *http://english.pravda.ru/columnists/2003/01/08/41736.html*

Wellman, B., and Haythornthwaite, C. (eds) (2002), *The Internet in Everyday Life*, Oxford: Blackwell.

Weseman, D. (1998), 'Unite, Don't Fight', *South African Bodyboarding Magazine* 7: 6.

West, N. (2000), *Kodak and the Lens of Nostalgia*, Charlottesville: University of Virginia Press.

Wickens, E. (2002), 'The Sacred and the Profane: A Tourist Typology', *Annals of Tourism Research* 29: 834–51.

Wildlife Preservation Trust International and Wildlife Conservation Society (1996), *Field Research and Community-Based Ecotourism Development: A Multidisciplinary Approach to the Conservation of the West Indian Manatee and its Marine Habitat in Belize, Central America* (proposal for a preliminary implementation grant), New York: Center for Environmental Research and Conservation.

Wilkinson, P. (1989), 'Strategies for Tourism in Island Microstates', *Annals of Tourism Research* 16: 153–77.

Williams, A., and Shaw, G. (1994), *Critical Issues in Tourism*, Oxford: Blackwell.

Williams, R. (1972), 'Ideas of Nature', in Benthall, J. (ed.), *Ecology: The Shaping Enquiry*, London: Longman.

Wittel, A. (2001), 'Towards a Network Sociality', *Theory, Culture and Society* 18: 51–76.

Wrathall, M., and Malpas, J. (eds) (2000), *Heidegger, Authenticity, and Modernity: Essays in Honour of Hubert L. Dreyfus*, vol. 1, Cambridge, MA: M.I.T. Press.

WTO (2002), *Yearbook of Tourism Statistics 2002*, Madrid: World Tourism Organization.

WTTC (2001a), Economic research: country league tables, *www.wttc.org/ecres/league.asp* (accessed 20 January 2002).

—— (2001b), Year 2001 TSA research summary and highlights: Cyprus, *www.wttc.org/ecres/a-cy.asp* (accessed 20 January 2002).

www.nzoom.com (2003) 'Discounts at the "SARS Hotel"', *http://onenews.nzoom.com/onenews_detail/0,1227,202844-1-456,00.html* (accessed 31 January 2004).

Yale, P. (1995), *The Business of Tour Operations*, Harlow: Longman.

Yelvington, K. (1995), *Producing Power: Ethnicity, Gender and Class in a Caribbean Workplace*, Philadelphia, PA: Temple University Press.

Ziehe, T. (1994), 'From Living Standard to Life Style', *Young* 2: 2–16.

Zimmerman, M.E. (2000), 'The End of Authentic Selfhood in the Postmodern Age?', in Wrathall, M. and Malpas, J. (eds) *Heidegger, Authenticity, and Modernity: Essays in Honour of Hubert L. Dreyfus*, vol. 1, Cambridge, MA: M.I.T. Press.

Zukin, S. (1991), *Landscapes of Power: From Detroit to Disney World*, Berkeley: University of California Press.

—— (1995), *The Cultures of Cities*, Oxford: Blackwell.

—— (2003), 'Home Shopping in the Global Marketplace', paper given to *Les sens de mouvement colloque*, Cerisy-la-salle, France, June.

Index

Note: Illustrations and tables are indicated by use of italics.

Adam, B. 196–7
adventure 71, 84; *see also* challenges of tourism;
    dangers of tourism; risks of tourism
Agatha 58
Agra Development Authority 104
Agra, Uttar Pradesh 103, 105–8, 109, 113–14; de-
    industrialization 110; historic sites 106; small
    traders 105, 107; *see also* Taj Mahal
'agrotourism' 25
Albertsen, N. 150
Alexander, Jacqui 19–20
Amigos Del Mar Dive Shop 38
Amigos Wreck dive 37–8
Amin, A., and Thrift, N. 185
Apocalypse Now 49
Appadurai, A. 16, 56–7, 187
Archaeological Society of India 109–10
Archaeological Survey of India 104, 111
architecture 59, 62, 147
Arellano, A. 36
Arlington National Cemetery 211
art exhibitions: *see* international travelling exhibitions
    (ITE)
Art Newspaper 91–2
Asian financial crisis 122
Asplund, J. 186, 187, 188
Augé, Marc 114
auratic objects 86
'authentic' experience 71–3, 76, 79, 85–6, 163–4
Ayia Napa, Cyprus 22, 27–8; club 'garage' scene
    24–5, 28, 30–1
Ayia Napa: Fantasy Island (Channel 4 series) 24, 28, 30–1

backpackers 104–5, 113
Bahamas 19–20
Barbados 16
Barber, Stephen: *Extreme Europe* 145, 148–9
Barcelona 131–41: development as a tourist venue
    132–4; Olympic Games 131, 134–6; place-myths
    136–7; post-Olympic development 136–41; and
    regeneration 131–2, 136–41; tourists and locals
    137–41; and urban tourism 135–6, *139*

'Barcelona model' 133–4, 135, 141
barrier reef 37–9
Bassett, C., and Wilbert, C. 172
Bath Spa 66
Baudrillard, J. 187
Baum, T.: *see* Conlin, M. and Baum, T.
Bauman, Z. 33–4, 56
The Beach 209
Beach Boys 48–9
beaches 2, 34, 36, 44–5, 181–2, 211–12; artificial 40
Bech, H. 150
Belfast: 'Troubles Tour' 214
Belize: Caribbean beach destination 34, 36; and
    ecotourism 32–41; environmental impact of
    ecotourism 36–40
Bell, C., and Lyall, J. 213
Bellah, R. 76, 77
Bellido, Efrain 69–70
Belmont, Frederick 60
Benjamin, W. 86
Bennett, Alan 61
Betty's Café Tea Rooms 60
Bingham, Hiram 67
Bishop, Elizabeth: 'The Burglar of Babylon'
    156–7
Black Orpheus 157
Blood of the Vikings (BBC series) 79
Bloom, W. 76
Blue Hole dive 37
Boase, J.: *see* Chen, W., Boase, J., and Wellman, B.
body, the 44–5, 70–3
Bohigas, Oriol 133
Böhme, G. 150
Bolter, Harold 195
Boo, E. 32
Bowcott, O. 105
Bradbury Space and Science Museum, USA 193–4
brands 4, 30, 59–60, 65–6
Bravo 20 194, 198–203; atomic history of Nevada
    198–9; bombing range 199–200; proposed
    National Park 192, 198; proposed Visitor Center
    and Museum 200–3, *201*, *202*

Brazil: hybridity 159; and international travelling exhibitions (ITE) 95; *see also favelas*
British Nuclear Fuels (BNFL) 192, 195
British Virgin Islands 18
Bryson, B. 58, 62
buildings 207; *see also* architecture
Bureau of Atomic Tourism 193–4
Buzard, J.: *The Beaten Track* 208

Cade, J. 58
Cadwalladr, C. 30
Camus, Albert 157
Camus, Marcel: *Black Orpheus* 157
capital: cultural 35–6, 99; social 189, 190; symbolic 208
Caribbean: Bahamas 19–20; Barbados 16; British Virgin Islands 18; Cuba 19, 23; 'globalization' of 15; investment in 16–17; islands 13, 22, 34, 36; Jamaica 20; migration 14–15; mobilities 14–20; Mustique 18; renaturalized 17; travel writing 17; as tropical paradise 13–21; *see also* Belize
Carroll, Corky 48
Castells, M. 57, 184, 188–9
Catalan style 133
Central Industrial Security Force (CISF) 110
challenges of tourism 4–5, 35, 38, 71–2; *see also* dangers of tourism; risks of tourism
Chan Chan rock art 46
Chaney, D. 65
Chek Lap Kok Airport, Hong Kong 121–2
Chen, W., Boase, J., and Wellman, B. 172
Chernobyl 196–7
Christie, Agatha 57–8
City of God (Cidade de Deus) 156
Claver, N.: *see* García, S., and Claver, N.
Clos, Joan 140
club 'garage' scene: Ayia Napa, Cyprus 24–5, 28, 30–1
Coleman, S., and Crang, M. 188
colonialism 17, 47, 120–1, 122
'commoditization' debate 79
Confederation of Indian Industry 111
conference tourism 9, 55, 57, 61, 62–4
Conlin, M., and Baum, T. 22
conservation 37–8, 68, 70, 104, 108–11; *see also* ecotourism; sustainable tourism
conservation holidays 33; *see also* ecotourism; sustainable tourism
Copenhagen: high-rise suburbs 144, *144*, 146
co-presences 189
coral reefs 37–9; artificial reef 38
corporeal performances 67, 71–3
corporeal travel 169, 170
cosmopolitanism 208
Crang, M. 177–8; *see also* Coleman, S., and Crang, M.
Cravens, Jen 72, 73
Cresswell, T. 212

Crouch, D. 57
cruise-ship industry 17
Cruz, Oscar 39
Cuba 19, 23
Cuban cars *19*
cultural capital 35–6, 99
cultural currency 99
'culture of nature' 73
Cuzco, Peru 74
cybertourism 170
Cyprus 22–31; Ayia Napa 22, 24–5, 27–8, 30–1; political and social structures 29; tourism development 26–8; tourism policy 24–9
Cyprus Tourism Organization (CTO) 25, 28–9

Dale, Dick 49
dangers of tourism 20, 196, 213; *see also* challenges of tourism; risks of tourism
'dark tourism' 149, 214
death: the ill and dying 211–13; places of 209–11
Delgado, M. 136
Denmark: high-rise suburbs 144, *144*, 146
developing countries: fetishization of 35–6
disease 20, 119–20, 126–9, 212
diving 34, 37–9
Dodge, M., and Kitchin, R. 175
Dos Santos, M. 94–5
drugs 16, 20, 155–6, 161
Dubai 181–91; aesthetic development of 185–6; airport 183, 184; economic development of 183–4; guest workers 184, 187; the Palm project 181–3, *182*, 185, 186–90; playscapes 187–90; technical development of 184–6; thematized environments 186–7, 188; women in 189, 190
Dubai Desert Classic 185
Dubai Shopping Festival 185
Duncan, J. 17

Eastern Europe: high-rise suburbs 143–4, 146
ecotourism 10, 32–41, 70–1, 208; environmental impact of 36–40; and global travel 33–6; and identity 33–4, 36; operators 37; *see also* sustainable tourism
Ecotourism Society 32
Edensor, T. 71, 73, 151, 152; *Tourists at the Taj* 103, 105, 109–10, 114
education and tourism 33; *see also* Bravo 20; museums; Sellafield Visitors Centre
el Raval, Barcelona 139–41
employment 4, 8; *see also* labour; work
environment 3, 36–40; *see also* ecotourism; nuclear tourism; sustainable tourism
environmental movement 33
Eriksson, Leif 78
'esoteric tourists' 73–7
ETA 213
extreme sports 212; *see also* Inca trail; surfing

Fainstein, S., and Judd, D. 135
Fallon Naval Air Station 198, 199
fantastic realism 85–7
Farmer, P. 212
favelas 156–60, 213; inhabitants of 156, 158–9, 160, 161, 164–5; representations of 155–60; Rio de Janeiro 159–61; Rio de Janeiro, São Conrado 160–1; Rocinha 160–5; tours 155, 158–65, 214
Featherstone, M. 187
fetishization of developing countries 35–6
Foley, M.: *see* Lennon, J., and Foley, M.
Forster, E. M. 208
France: high-rise suburbs 143, 146; Paris 149, 153–4, 158, 214
Free Zones 183–4
Freud, S. 206
Friends of the Earth 197
Fujimori, Alberto 69

García, S. 134; and Claver, N. 135, 140
Garland, Alex: *The Beach* 209
gated communities 160, 190
Gaudi, Antoni 135
'gaze' of the tourist 70–3, 194
genealogy 82–4, 86; genealogical tourism 87
'glance' of the tourist 65
globalization 3, 4, 8–10; of culture 15, 56–7; economic 16–17, 119; and ecotourism 33–6; of events 9, 94, 134–6; of places 2, 56–7, 134, 169–70, 182, 206
'glocalization' 103–4, 111–15
Gothic Quarter, Barcelona 139
Graham, S., and Marvin, S. 190
Griffin, Rick 48
'Ground Zero' 214
Guggenheim Museum 94
guides 37–9, 74–5

La Haine 149
Harrogate 55–66; annual events 61; nightlife 64–5; conference tourism 61–6; heritage tourism 57–61; brands of 59–60, 65–6; Royal Baths 63, 65
Harrogate Advertiser 62, 64
Harrogate Borough Council 59
Harrogate District Guide 2003 55
Harrogate International Centre 62, 63, 64
Harrogate Spa Water 65–6
Harvey, D. 61, 183
Hattersley, Roy 58
Haussmann, Georges Eugène, Baron 158
Havana, Cuba 19
Hawaii 48
hedonism 17–18, 23, 36; *see also* playfulness; sex tourism
Heelas, P. 77
Heidegger, M. 150

heritage tourism 2, 18–19, 57–61, 68–9, 78–9, 85–7
Herriot, James 59
Hetherington, K. 57
high-rise suburbs 143–54, *144*; aesthetic judgements on 146–7; atmosphere of 150–1; and boredom 147–8; and the city centre 153–4; and excitement 148–9; and nostalgia 147; place-myths 144, 147; and the sublime 151; suburban tourism 143–5, 153; and tourist performance 151–3; tourists' attitude to inhabitants 153–4; tourists' interpretations of 145–7
Hiroshima 192, *210*; memorial 193, *193*
Hol Chan Marine Reserve 37
Holidays in the Danger Zone (BBC TV series) 212
Hong Kong 119–30; Chinese mainland visitors to 124–6, 129; colonial image of 120–1, 122; Disneyland 124–5; as East-West playscape 120–6; handover 121–2; Harbour *123*; and 'otherness' 120, 121; rebranded as 'City of Life' 123–5; and SARS 126–9, *127*; as virtual West 124–6; visitor arrivals, annual *122*; visitor arrivals, monthly *128*; Western tourists 124
Hong Kong Harbour *123*
Hong Kong Tourist Association (HKTA) 120, 123, 124
'host city' sites 9
Hoving, Thomas 90–1
Huang, K. 97
Huizinga, J. 50

Ibiza 23
Icelandic sagas 82–3
Inca trail 70–3
Incas 67, 73; and mythology 74
independent travel 33, 34–5
Indian Supreme Court 108, 110
Inside Out (BBC TV programme) 64–5
Institute for Transportation and Development Policy 110
interactive travel 172, 175–6, 178
International Council of Monuments and Sites (ICOMOS) 69–70
international travelling exhibitions (ITE) 90–101; characteristics of 98–101; emergence of 90–5; repertoire 94; most popular exhibitions 91, *92*, *93*; in Taiwan 97–8; visitor numbers 94–5
internet 9; *see also* websites
Internet City 184
islands 2, 13–21, 32, 34, 36; *see also* Palm project

Jagmohan, Shri 111
Jamaica 20
Jamuna River, India 109, 111
Jan & Dean 49
Judd, D.: *see* Fainstein, S., and Judd, D.
Julier, G. 133

Kahanamoku 47, 48
Kaohsiung Fine Art Museum, Taiwan (KFAM) 97
Kaplan, C. 15–16
Kashmir, dispute over 105, 107
Kassovitz, Mathieu: *La Haine* 149
King, R. 22–5
Kitchin, R.: *see* Dodge, M., and Kitchin, R.
Kluge, A.: *see* Negt, O., and Kluge, A.

la Ribera, Barcelona 139–41
labour 15–16; *see also* employment; work
Lake District 195
land 29, 206–7
lap-dancing 64
Las Ramblas, Barcelona 137
Lash, S. 50, 185
Le Corbusier 156, 157
Leith, William 66
Lennon, J., and Foley, M. 149
Lévi-Strauss, C. 158
Lewis, N. 71
lifestyle 4, 136
Lodge, David: *Paradise News* 206
Löfgren, O. 151
logos 4; *see also* brands
Lone Rock 199–200
Los Angeles ghettos 149
Lury, C. 60, 99
Lyall, J.: *see* Bell, C., and Lyall, J.

MacCannell, D. 76, 151, 163–4
Machu Picchu 67–77, 207, 213; development project
    69–70; Inca trail 70–3; and esoteric tourists
    73–7; World Heritage Site 68–70
Macnaghten, P., and Urry, J. 73
Maktoum, Mohammed bin Rashid Al 181, 184
Malbon, B. 64
Mallorca 23
Mann, T. 213; *Death in Venice* 212
Maragall, Pasqual 133, 134
Maremagnum, Barcelona 137–8, *138*, 140
Marvin, S.: *see* Graham, S., and Marvin, S.
Maspero, François: *Roissy Express* 145, 147–8, 149, 152
Massey, D. 57, 61
Maurer, B. 17, 18
McCalla, W. 40
McTiernan, John 88
Mediterranean islands 22
Meier, Richard 139
Melanesian islands 22
Messel, Oliver 18
Metropolitan Museum, New York 90–1
Military Land Withdrawal Act 200
miscegenation 159
Misrach, Richard 199–203; *Bravo 20: The Bombing of
    the American West* 200–3
Molotch, H. 60, 209

Mowforth, M., and Munt, I. 33, 208
Munt, I. 33, 35–6; *see also* Mowforth, M., and
    Munt, I.
Murphy, Bill 65
Museum of Modern Art, Barcelona 139–40
museums: capital 98; cooperation between 101; and
    funding 91, 96; hierarchy in 98, 99–100; most
    popular exhibitions 91, *92*, *93*; practices of
    100; in Taiwan 90, 95–8; *see also* international
    travelling exhibitions (ITE); Viking Ship
    Museum, Roskilde, Denmark
Mustique 18
Mykonos 23
'mystic gaze' 74–5, 76

Nagasaki: memorial 193
National Museum of History, Taiwan 97
National Palace Museum, Taiwan (NPM) 95, 96, 97
National Tourism Corporation, Peru 69
nature tourism: *see* ecotourism
Neesam, M. 62
Negt, O., and Kluge, A. 85
Nellis Air Force Range 199
network society 184, 189
Nevada Test Site 198
New Age movement: and Machu Picchu 74–7
New York City 15
New York Magazine 90–1
New Zealand 213
nuclear industry 195–7, 199
nuclear tourism 192–203; Bravo 20 192, 194,
    198–203; definition of 193–4; Sellafield
    Visitors Centre 192, 195–8, 203
Nuclear Waste Technical Review Board (NWTRB)
    199
Nunez, Daniel 38–9

Observer Food Monthly 66
offshore banking 16–17
Old City, Barcelona 138–40
Olympic Games: Barcelona 131, 134–6
Orbitz 169
orientalist tourism 18

package-holiday industry 17
Paiute nation 199
Palm project 181–3, *182*, 185, 186–90
Paphos, Cyprus 24, 27; Harbour *25*
Paris: no-go areas 149; redesigned 158; suburbs
    153–4, 214
Parr, Martin: *Boring Postcards* 147
'patriotic tourism' 214
Pattullo, Polly 13
performances 1, 4–5, 6–8, 205–9; Ayia Napa club scene
    28; corporeal 67, 71–3; ecotourist 38; heritage
    57–61; high-rise suburb 151–3; surfing 45–6
Peru: *see* Machu Picchu

photography 151–2, 177–8, 207
picturesque, the 17
'placelessness' 56, 65, 169
place-myths 57–66, 136–7, 140, 144, 147
places: and death 209–11, 214; and globalization 56–7, 169–70, 182, 206; and mobility 5–10, 14–20, 208; and play 5–10, 205–10; and performances 1, 4–5, 6–8, 205, 206, 209; and playfulness 1, 23, 186–8; virtual 4
playfulness 1, 23, 186–8; see also hedonism; sex tourism
playscapes 120–6, 187–90
political struggles 7
'post-tourists' 70
postmodernism 187
Prescott, John 195
preservation: see conservation; ecotourism; sustainable tourism
Preston-Whyte, R. 45–6
Protaras, Cyprus 27

'quality' tourism 27

radiation 197–8
radiation contamination 193, 198, 199
Radioactive Waste Executive, NIREX 197
Raitz, Vladimir 22
refugees 3
Richards, Julian: Blood of the Vikings 79
Rio de Janeiro: see favelas
risks of tourism 5, 196, 212; of surfing 46; see also challenges of tourism; dangers of tourism
rite of passage 71–2
Roche, M. 134
Rocinha, Rio de Janeiro 160–5
rock art 46
Rojek, C. 149
Royal Baths, Harrogate 62, 63, 65

San Pedro (Ambergris Caye), Belize 37–8, 40
Sánchez Taylor, J. 20
sand piracy 40
São Conrado, Rio de Janeiro 160–1
SARS (Severe Acute Respiratory Syndrome) 119–20, 126–9, 212
Schultz, P.: 1000 Places to See Before You Die 208
Science Museum, London 196
seasonal tourism 27
security 189–90, 198, 214
self, the 35–6, 50–1, 76
Sellafield Visitors Centre 192; 'The Mighty Atom' exhibit 195; nuclear tourism 194–8; 'Sparking Reaction' exhibit 195–8
Sellars, A. 30
service work: see labour; employment; work
Severe Acute Respiratory Syndrome (SARS) 119–20, 126–9, 212

Severson, John 48
sex tourism 19–20; see also hedonism; playfulness
Seychelles 23
shamans 74
Shark Ray Alley 37–8, 39
Shaw, G.: see Williams, A., and Shaw, G.
Sheller, M. 36
smuggling 16
snorkelling 37–8
social capital 189, 190
spa towns 211; see also Harrogate
space travel 212–13
special-interest activities 33
spirituality 10; see also New Age Movement; self, the
Squire, S. 56
Strachan, I. 13–14
Subiros, P. 133
sublime, the 151
suburban tourism 143–5, 153; see also favelas; high-rise suburbs
Sunburn (BBC series) 24
Supersonic Operations Area 200
surfing 44–51; board design and technology 48, 49, 50; conflicts 46; culture 45–6; and diplomacy 49; and Edwardian globalization 47; images of 48, 50; magazines 45, 46–7, 48; origins of 46–7; as play 50; professionalism 48; as social performance 45–6; 'surf' music 49
Surfing (magazine) 45, 48
sustainable tourism 24, 25, 71
symbolic capital 208
'symbolic economy' 134–5

Taipei Fine Art Museum (TFAM), Taiwan 95–7
Taiwan: and International travelling exhibitions (ITE) 90, 95–8; museums 95–8; social change 96; structural transformations 95
Taj Ganj 104–5, 107–8; small traders 105, 107, 113–14
Taj Mahal 103–15, 207; admission charges 104–8; Agra, Uttar Pradesh 103, 105–8, 109, 110, 113–14; backpackers 104–5, 113; conservation 108–11, 112–13; development of site 104, 105, 106, 108–11, 113; domestic tourists 105, 107, 113; global tourist flows 113–14; and 'glocalization' 103–4, 111–15; opening hours 106; regulation 114–15; security 107, 110, 112–13
Taj Mahal Cycle Taxi Improvement Project 110
Taj National Park 109
Tarragona, Spain 193
tax havens 16–17
taxation on tourism 17
Taylor's of Harrogate 59–60
technologies 15–16, 48, 49, 50, 119, 184–6; see also websites
terrorism 2, 105, 119–20, 213–14; threat of 107, 112
Thermae 66
The 13th Warrior 88

Thomas Cook: 'Circular Tour' 169
Thomas-Hope, Elizabeth 14
Thompson, E.P.: *Protest and Survive* 196
THORP reprocessing plant 195
Thrift, N.: *see* Amin, A., and Thrift, N.
Tibbets, Paul; and the *Emnola Gay* bombing 192
Toledo, Alejandro 69
Tolkien, J.R.R. 86
tour guides 38–9, 74–5
Tour Guides Association, San Pedro 38
tour operators 27, 28–9, 30, 37, 39, 105–6
tourism mobilities 1–10, 15–16, 205–6: *see also* 'agro-
    tourism'; conference tourism; cybertourism; 'dark
    tourism'; heritage tourism; nuclear tourism;
    orientalist tourism; 'patriotic' tourism; 'quality'
    tourism; seasonal tourism; sex tourism; suburban
    tourism; sustainable tourism; urban tourism
travel: corporeal 169, 170; independent 33, 34–5;
    interactive 172, 175–6, 178; space 212–13;
    virtual 170–8
Tung Chee Hwa 122

UNESCO 69–70, 108, 111
urban protests 8
urban tourism 135–6; *see also favelas*; high-rise
    suburbs; suburban tourism
Urry, J. 33, 56, 62, 184, 186, 189; and 'glocalization'
    103; 'the gaze' 194; *see also* Macnaghten, P., and
    Urry, J.
utopian architecture 147

Victoria Gardens Shopping Centre, Harrogate 62
Viking Ship Museum, Roskilde, Denmark 79, 84–7;
    Activity Room *81*, 83; map *80*; replica ship *81*,
    82, 85–7; ship wrecks *80*, 82, 85–6
Vikings 78–88; culture 87–8; fantastic elements
    83–4, 85–7; and genealogy 82–4, 87; iconog-
    raphy 87; mobilities 87–8; in popular culture
    83
violence 148–9, 155–6, 158, 159
virtual places 4
virtual travel 170–8

Wallpaper 147
Walt Disney Company (Asia Pacific) 124–5
'War on Drugs' 20
Wardle, Huon 14
waves: *see* surfing
websites 9, 10, 169–72; and corporeal travel 169,
    170; daily reports *176*; home and away 171–2;
    and photography 177–8; real and virtual
    175–8; 'real time' 169; round-the-world
    169–72; and virtual travel 170–8; as work
    173–5; work and leisure 172–3, 175
Welles, Orson 157
Wellman, B.: *see* Chen, W., Boase, J., and
    Wellman, B.
Wendover Air Force Base 198
West Edmonton Mall: Paris theme 188
Western Europe: high-rise suburbs 144
Western, Rachel 197
Western Shoshone nation 199
Wilbert, C.: *see* Bassett, C., and Wilbert, C.
Williams, A., and Shaw, G. 62
Williams, Raymond 207
Windscale: *see* Sellafield Visitors Centre
Wittel, A. 172, 188–9
Wong Tai Sin Temple 120
Wordsworth, W.: *The Brother* 208
work, service-sector 18, 60–1, 70; *see also* employ-
    ment; labour
working-class communities 17–18, 136, 137–41, 152
World Heritage Centre 69–70
World Heritage Convention 68, 108, 112
World Heritage Sites 67, 68–70, 108, 110–11
World Trade Centre 137: *see also* terrorism

'Yardie' gangs 20
Yaxk'in, Aluna Joy 75–6
Yorkshire Dales 59
youth gangs 148–9
Yucca Mountain 199

Zukin, S. 7